Edward R. Miller-Jones (Hrsg.)

Space Tourism

Edward R. Miller-Jones (Hrsg.)

Space Tourism

Traveling beyond Earth

Fastbook Publishing

Imprint

Publisher:
Fastbook Publishing is a trademark of
VDM Publishing House Ltd.,17 Rue Meldrum, Beau Bassin,1713-01 Mauritius
Email: info@vdm-publishing-house.com
Website: www.vdm-publishing-house.com

Published in 2010

Printed in: U.S.A., U.K., Germany. This book was not produced in Mauritius.

ISBN: 978-613-0-10603-4

Introduction

The unstoppable wish to reach the stars

Exploring outer space has been a man's urge throughout the history of human kind. Ancient civilizations in China, Egypt, Greece and Babylonia devoted their attention to observing the stars, the Moon, lunar eclipses and various other space-related objects. Since then one has always wanted to actually experience what it feels like being beyond earth. Thus the era of space exploration began.

The Soviet Union launched the first Earth-orbiting artificial satellite - Sputnik 1 - in 1957, and Yuri Gagarin became the first man to travel into outer space in 1961[1]. The United States replied with the Apollo 11 mission that landed the first humans on the Moon in 1969. All these are considered to be significant milestones in the field of space exploration.

Today traveling into outer space is not the privilege of just highly qualified astronauts. Ever since space travels became frequent and the International Space Station (a research facility where experiments in biology, physics, astronomy and meteorology are being conducted) began operating, there have been several people whose determination to experience what it feels like to be in outer space has proved to be unstoppable. The pioneer in the field of space tourism is Dennis Tito (an American engineer) who was the first man traveling into space and paying for it with his own money (the ``space ticket`` reportedly cost him 20 million US dollars[2]). Tito himself, however, claims it was worth it, since space traveling had been his dream for decades.

Soon enough several other adventurers followed Tito's example. Mark Shuttleworth (an entrepreneur from South Africa), Gregory Olsen, Anousheh Ansari (she became the first Muslim woman and the first Iranian in space[3]), Charles Simonyi (a Hungarian-born American computer software executive), Richard Allen Garriott (a video-game developer whose flight reportedly cost a staggering 30 million USD[4]) and Guy Laliberté (the CEO of Cirque du Soleil) all have used the opportunity provided by Space Adventures, Ltd. (a

1Encyclopedia Britannica Online: Science & Technology: Yury Alekseyevich Gagarin, (accessed: 06/2010)
2 BBC News Online: Sci/Tech: Profile: Tito the spaceman, (date: 28/04/2001)
3 The Independent Online: News: Science: Pride in space as Iran cheers first Muslim's journey to the stars, (date: 19/09/2006)

United States-based space tourism company). Each of the aforementioned people successfully reached the International Space Station and stayed there for several days, helping out official astronauts, recording their experiences and enjoying the sight of the Earth from space.

These sought-after space adventures have created a whole business industry, focusing on providing space-related experiences to individuals who are not astronauts and thus have to find their own way how to fulfill the dream of traveling beyond Earth. Several companies have been established with the goal to providing space tourism-related services, among them the most notable are the above mentioned Space Adventures, the ever-innovative Sir Richard Branson's Virgin Galactic and Bigelow Aerospace (a Nevada-based company that plans to launch real modules into space in order to create the solar system's first private space station, one that could accommodate 36 people at a time).

This book gives an exciting and comprehensive insight on every important issue regarding the emerging field of space tourism. Here you will find chapters on topics like outer space in general, the human spaceflight (and the corresponding fields of space medicine and the space adaptation syndrome), space tourism and spaceflight participants, the Soyuz spacecraft (its models have been used by every space tourist), the International Space Station and all of the already mentioned people who have traveled into outer space as tourists. This book concludes with chapters on the emerging business ventures determined to work in the field of space travels.

4 The Houston Chronicle Online: $30 million buys Austin resident a ride on Soyuz mission, (date: 10/10/2008)

Contents

Articles

References

Article Licenses

Outer space

Outer space (often simply called **space**) is the void that exists beyond any celestial body including the Earth.[1] It is not completely empty (i.e. a perfect vacuum), but contains a low density of particles, predominantly hydrogen plasma, as well as electromagnetic radiation, magnetic fields, and neutrinos. Theoretically, it also contains dark matter and dark energy.

The boundaries between the Earth's surface and outer space, at the Kármán line, 100 km (62 mi) and exosphere at 690 km (430 mi).

Discovery

In 350 BC, Greek philosopher Aristotle suggested that *nature abhors a vacuum*, a principle that became known as the *horror vacui*. Based on this idea that a vacuum could not exist, it was widely held for many centuries that space could not be empty.[2] As late as the seventeenth century, the French philosopher René Descartes argued that the entirety of space must be filled.[3] It became known to Galileo Galilei that air had weight and so was subject to gravity. He also demonstrated that there was an established force that resisted the formation of a vacuum. However, it would remain for his pupil Evangelista Torricelli to create an apparatus that would produce a vacuum. At the time this experiment created a scientific sensation in Europe. The French mathematician Blaise Pascal reasoned that if the column of mercury was suspended by air then the column ought to be shorter at higher altitude. His brother in law, Florin Périer, repeated the experiment on the Puy-de-Dôme mountain in central France and found that the column was shorter by three inches. This decrease in pressure was further demonstrated by carrying a half-full balloon up a mountain and watching it gradually inflate, then deflate upon descent. These and other experiments were used to overthrow the principle of *horror vacui*.[4]

Further work on the physics of the vacuum was performed by Otto von Guericke. He correctly noted that the atmosphere of the Earth surrounds the planet like a shell, with the density gradually declining with altitude. He concluded that there must be a vacuum between the Earth and the Moon.[5]

Early speculations as to the infinite dimension of space was performed in the sixteenth century by the Italian philosopher Giordano Bruno. He extended the Copernican heliocentric cosmology to the concept of an infinite universe that is filled with a substance he called aether, which did not cause resistance to the motions of heavenly bodies.[6] English philosopher William Gilbert arrived at a similar conclusion, arguing that the stars are visible to us only because they are surrounded by a thin aether or a void.[7] This concept of an aether originated with ancient Greek philosophers, including Aristotle, who conceived of it as the medium through which the heavenly bodies moved.[8]

The concept of a universe filled with a luminiferous aether remained in vogue among some scientists up until the twentieth century. This form of aether was viewed as the medium through which light could propagate. In 1887, the Michelson-Morley experiment was carried out as an attempt to detect the Earth's motion through this medium by looking for changes in the speed of light based on the direction of the planet's motion. However, the null result indicated something was wrong with the concept. Since then the idea of the luminiferous aether had essentially been abandoned, to be replaced by Albert Einstein's theory of special relativity. The latter held that the speed of light is a constant in a vacuum, regardless of the observer's motion or frame of reference.[9] [10]

The first professional astronomer to support the concept of an infinite universe was the Englishman Thomas Digges in 1576.[11] However, the true scale of the universe remained unknown until the first successful measurement of the distance to a nearby star was performed in 1838 by the German astronomer Friedrich Bessel. He showed that the star 61 Cygni had a parallax of just 0.31 arcseconds (compared to the modern value of 0.287″). This corresponded to a distance of over 10 light years.[12] The distance scale to the Andromeda galaxy was determined in 1923 by American astronomer Edwin Hubble when he measured the brightness of cepheid variables within that galaxy. This established that the Andromeda galaxy, and by extension all galaxies, lay well outside the Milky Way.[13]

The modern concept of outer space is based upon the Big Bang cosmology, which was first proposed in 1931 by the Belgian physicist Georges Lemaître. This theory holds that the observable universe originated from a very compact form that has since undergone continuous expansion. Matter that remained following the initial expansion has since undergone gravitational collapse to create stars, galaxies and other astronomical objects, leaving behind a deep vacuum that forms what is now called outer space.[14]

The term *outer space* was first recorded by the English poet Lady Emmeline Stuart-Wortley in her poem "The Maiden of Moscow" in 1842,[15] and later popularised in the writings of HG Wells in 1901.[16] The shorter term *space* is actually older, first used to mean the region beyond Earth's sky in John Milton's *Paradise Lost* in 1667.[17]

Environment

Outer space is the closest natural approximation of a perfect vacuum. It has effectively no friction, allowing stars, planets and moons to move freely along ideal gravitational trajectories. However, even in the deep vacuum of intergalactic space there are still a few hydrogen atoms per cubic meter.[18] By comparison, the air we breathe contains about 10^{25} molecules per cubic meter.[19] The sparse density of matter in outer space means that electromagnetic radiation can travel great distances without being scattered; the mean free path for a photon in intergalactic space is about 10^{23} km, or 10 billion light years.[20] The deep vacuum of space could make it an attractive environment for certain industrial processes, for instance those that require ultraclean surfaces.[21]

Stars, planets, asteroids and moons keep their atmospheres by gravitational attraction, and as such, atmospheres have no clearly delineated boundary: the density of atmospheric gas simply decreases with distance from the object. The Earth's atmospheric pressure drops to about 1 Pa at 100 kilometres (62 mi) of altitude. This is known as the Kármán line, a common definition of the boundary with outer space. Beyond this line, isotropic gas pressure rapidly becomes insignificant when compared to radiation pressure from the Sun and the dynamic pressure of the solar wind, so the definition of pressure becomes difficult to interpret. The thermosphere in this range has large gradients of pressure, temperature and composition, and varies greatly due to space weather. Astrophysicists prefer to use number density to describe these environments, in units of particles per cubic centimetre.

Temperature

All of the observable Universe is filled with photons that were created during the Big Bang, which is known as the cosmic microwave background radiation (CMB). There is quite likely a correspondingly large number of neutrinos called the cosmic neutrino background. The current black body temperature of this photon radiation is about 3 K (−270 °C; −454 °F). Some regions of outer space can contain highly energetic particles that have a much higher temperature than the CMB.

Effect on human bodies

Contrary to popular belief,[22] a person suddenly exposed to the vacuum would not explode, freeze to death or die from boiling blood. Air would immediately leave the lungs due to the enormous pressure gradient. Any oxygen dissolved in the blood would empty into the lungs to try to equalize the partial pressure gradient. Once the deoxygenated blood arrived at the brain, death would quickly follow.

Humans and animals exposed to vacuum will lose consciousness after a few seconds and die of hypoxia within minutes. Blood and other body fluids do boil when their pressure drops below 6.3 kPa, the vapor pressure of water at body temperature.[23] This condition is called ebullism. The steam may bloat the body to twice its normal size and slow circulation, but tissues are elastic and porous enough to prevent rupture. Ebullism is slowed by the pressure containment of blood vessels, so some blood remains liquid.[24] [25] Swelling and ebullism can be reduced by containment in a flight suit. Shuttle astronauts wear a fitted elastic garment called the Crew Altitude Protection Suit (CAPS) which prevents ebullism at pressures as low as 2 kPa.[26] Water vapor would also rapidly evaporate off from exposed areas such as the lungs, cornea of the eye and mouth, cooling the body. Rapid evaporative cooling of the skin will create frost, particularly in the mouth, but this is not a significant hazard. Space may be cold, but it's mostly vacuum and transfers heat ineffectively; as a result the main temperature regulation concern for space suits is how to get rid of naturally generated body heat.

Cold or oxygen-rich atmospheres can sustain life at pressures much lower than atmospheric, as long as the density of oxygen is similar to that of standard sea-level atmosphere. The colder air temperatures found at altitudes of up to 3 kilometres (1.9 mi) generally compensate for the lower pressures there.[23] Above this altitude, oxygen enrichment is necessary to prevent altitude sickness, and spacesuits are necessary to prevent ebullism above 19 kilometres (12 mi).[23] Most spacesuits use only 20 kPa of pure oxygen, just enough to sustain full consciousness. This pressure is high enough to prevent ebullism, but simple evaporation of blood can still cause decompression sickness and gas embolisms if not managed.

Rapid decompression can be much more dangerous than vacuum exposure itself. Even if the victim does not hold his breath, venting through the windpipe may be too slow to prevent the fatal rupture of the delicate alveoli of the lungs.[23] Eardrums and sinuses may be ruptured by rapid decompression, soft tissues may bruise and seep blood, and the stress of shock will accelerate oxygen consumption leading to hypoxia.[27] Injuries caused by rapid decompression are called barotrauma. A pressure drop as small as 13 kPa, which produces no symptoms if it is gradual, may be fatal if it occurs suddenly.[23]

Boundary

There is no clear boundary between Earth's atmosphere and space, as the density of the atmosphere gradually decreases as the altitude increases. There are several designated scientific boundaries, namely:

- The Fédération Aéronautique Internationale has established the Kármán line at an altitude of 100 kilometres (62 mi) as a working definition for the boundary between aeronautics and astronautics. This is used because above an altitude of roughly 100 km, as Theodore von Kármán calculated, a vehicle would have to travel faster than orbital velocity in order to derive sufficient aerodynamic lift from the atmosphere to support itself.
- The United States designates people who travel above an altitude of 50 miles (80 km) as astronauts.
- NASA's mission control uses 76 miles (122 km) as their re-entry altitude, which roughly marks the boundary where atmospheric drag becomes noticeable, (depending on the ballistic coefficient of the vehicle), thus leading shuttles to switch from steering with thrusters to maneuvering with air surfaces.

In 2009, scientists at the University of Calgary reported detailed measurements with an instrument called the Supra-Thermal Ion Imager (an instrument that measures the direction and speed of ions), which allowed them to determine that space begins 118 kilometres (73 mi) above Earth. The boundary represents the midpoint of a gradual transition over tens of kilometers from the relatively gentle winds of the Earth's atmosphere to the more violent flows of charged particles in space, which can reach speeds well over 600 miles per hour (1000 km/h).[28] [29]

This was only the second time that direct measurements of charged particle flows have been conducted at this region, which is too high for balloons and too low for satellites. It was however the first study to include all the relevant elements for this kind of determination – for example, the upper atmospheric winds.

The instrument was carried by the JOULE-II rocket on January 19, 2007, and traveled to an altitude of about 124 miles (200 km). From there it collected data while it was moving through the "edge of space".[28]

Legal status

The Outer Space Treaty provides the basic framework for international space law. This treaty covers the legal use of outer space by nation states, and includes in its definition of *outer space* the Moon and other celestial bodies. The treaty states that outer space is free for all nation states to explore and is not subject to claims of national sovereignty. It also prohibits the deployment of nuclear weapons in outer space. The treaty was passed by the United Nations General Assembly in 1963 and signed in 1967 by the USSR, the United States of America and the United Kingdom. As of January 1, 2008 the treaty has been signed by 98 states and ratified by an additional 27 states.[30]

Between 1958 and 2008, outer space has been the subject of multiple resolutions by the United Nations General Assembly. Of these, more than 50 have been concerning the international co-operation in the peaceful uses of outer space and preventing an arms race in space.[31] Four additional space law treaties have been negotiated and drafted by the UN's Committee on the Peaceful Uses of Outer Space. The 1979 Moon Treaty turned the jurisdiction of all heavenly bodies (including the orbits around such bodies) over to the international community. However, this treaty has not been ratified by any nation that currently practices manned spaceflight.[32]

Space versus orbit

To perform an orbit, a spacecraft must travel faster than a sub-orbital spaceflight. A spacecraft has not entered orbit until it is traveling with a sufficiently great horizontal velocity such that the acceleration due to gravity on the spacecraft is less than or equal to the centripetal acceleration being caused by its horizontal velocity (see circular motion). So to enter orbit, a spacecraft must not only reach space, but must also achieve a sufficient orbital speed (angular velocity). For a low-Earth orbit, this is about 7900 m/s (28400 km/h; 17700 mph); by contrast, the fastest airplane speed ever achieved (excluding speeds achieved by deorbiting spacecraft) was 2200 m/s (7900 km/h; 4900 mph) in 1967 by the North American X-15.[33]

Konstantin Tsiolkovsky was the first person to realize that, given the energy available from any available chemical fuel, a several-stage rocket would be required. The escape velocity to pull free of Earth's gravitational field altogether and move into interplanetary space is about 11000 m/s (39600 km/h; 24600 mph) The energy required to reach velocity for low Earth orbit (32 MJ/kg) is about twenty times the energy required simply to climb to the corresponding altitude (10 kJ/(km·kg)).

There is a major difference between sub-orbital and orbital spaceflights. The minimum altitude for a stable orbit around Earth (that is, one without significant atmospheric drag) begins at around 350 kilometres (220 mi) above mean sea level. A common misunderstanding about the boundary to space is that orbit occurs simply by reaching this altitude. Achieving orbital speed can theoretically occur at any altitude, although atmospheric drag precludes an orbit that is too low. At sufficient speed, an airplane would need a way to keep it from flying off into space, but at present, this speed is several times greater than anything within reasonable technology.

A common misconception is that people in orbit are outside Earth's gravity because they are "floating". They are floating because they are in "free fall": they are accelerating toward Earth, along with their spacecraft, but are simultaneously moving sideways fast enough that the "fall" away from a straight-line path merely keeps them in orbit at a constant distance above Earth's surface. Earth's gravity reaches out far past the Van Allen belt and keeps the Moon in orbit at an average distance of 384403 kilometres (238857 mi).

Regions

Space is not a perfect vacuum: its different regions are defined by the various atmospheres and "winds" that dominate within them, and extend to the point at which those winds give way to those beyond. Geospace extends from Earth's atmosphere to the outer reaches of Earth's magnetic field, whereupon it gives way to the solar wind of interplanetary space. Interplanetary space extends to the heliopause, whereupon the solar wind gives way to the winds of the interstellar medium. Interstellar space then continues to the edges of the galaxy, where it fades into the intergalactic void.

Geospace

Geospace is the region of outer space near the Earth. Geospace includes the upper region of the atmosphere, as well as the ionosphere and magnetosphere. The Van Allen radiation belts also lie within the geospace. The region between Earth's atmosphere and the Moon is sometimes referred to as **cis-lunar space**.

Aurora australis observed by *Discovery*, May 1991.

Although it meets the definition of outer space, the atmospheric density within the first few hundred kilometers above the Kármán line is still sufficient to produce significant drag on satellites. Most artificial satellites operate in this region called low earth orbit and must fire their engines every few days to maintain orbit. The drag here is low enough that it could theoretically be overcome by radiation pressure on solar sails, a proposed propulsion system for interplanetary travel.

Geospace is populated by electrically charged particles at very low densities, the motions of which are controlled by the Earth's magnetic field. These plasmas form a medium from which storm-like disturbances powered by the solar wind can drive electrical currents into the Earth's upper atmosphere.

During geomagnetic storms two regions of geospace, the radiation belts and the ionosphere, can become strongly disturbed. These storms increase fluxes of energetic electrons that can permanently damage satellite electronics, disrupting telecommunications and GPS technologies, and can also be a hazard to astronauts, even in low-Earth orbit. They also create aurorae seen near the magnetic poles.

Geospace contains material left over from previous manned and unmanned launches that are a potential hazard to spacecraft. Some of this debris re-enters Earth's atmosphere periodically.

The absence of air makes geospace (and the surface of the Moon) ideal locations for astronomy at all wavelengths of the electromagnetic spectrum, as evidenced by the spectacular pictures sent back by the Hubble Space Telescope, allowing light from about 13.7 billion years ago — almost to the time of the Big Bang — to be observed.

The outer boundary of geospace is the interface between the magnetosphere and the solar wind. The inner boundary is the ionosphere.[34] Alternately, geospace is the region of space between the Earth's upper atmosphere and the outermost reaches of the Earth's magnetic field.[35]

Interplanetary

Interplanetary space, the space around the Sun and planets of the Solar System, is the region dominated by the interplanetary medium, which extends out to the heliopause where the influence of the galactic environment starts to dominate over the magnetic field and particle flux from the Sun. Interplanetary space is defined by the solar wind, a continuous stream of charged particles emanating from the Sun that creates a very tenuous atmosphere (the heliosphere) for billions of miles into space. This wind has a particle density of 5–10 protons/cm^3 and is moving at a velocity of 350–400 km/s.[36] The distance and strength of the heliopause varies depending on the activity level of the solar wind.[37] The discovery since 1995 of extrasolar planets means that other stars must possess their own interplanetary media.[38]

The volume of interplanetary space is an almost pure vacuum, with a mean free path of about one astronomical unit at the orbital distance of the Earth. However, this space is not completely empty, and is sparsely filled with cosmic rays, which include ionized atomic nuclei and various subatomic particles. There is also gas, plasma and dust, small meteors, and several dozen types of organic molecules discovered to date by microwave spectroscopy.[39]

Interplanetary space contains the magnetic field generated by the Sun.[36] There are also magnetospheres generated by planets such as Jupiter, Saturn and the Earth that have their own magnetic fields. These are shaped by the influence of the solar wind into the approximation of a teardrop shape, with the long tail extending outward behind the planet. These magnetic fields can trap particles from the solar wind and other sources, creating belts of magnetic particles such as the Van Allen Belts. Planets without magnetic fields, such as Mars and Mercury, but excluding Venus, have their atmospheres gradually eroded by the solar wind.

Interstellar

Interstellar space is the physical space within a galaxy not occupied by stars or their planetary systems. The interstellar medium resides—by definition—in interstellar space.

Intergalactic

Intergalactic space is the physical space between galaxies. Generally free of dust and debris, intergalactic space is very close to a total vacuum. The space between galaxy clusters, called the voids, is probably nearly empty. Some theories put the average density of the Universe as the equivalent of one hydrogen atom per cubic meter.[40] [41] The density of the universe, however, is clearly not uniform; it ranges from relatively high density in galaxies (including very high density in structures within galaxies, such as planets, stars, and black holes) to conditions in vast voids that have much lower density than the universe's average.

Surrounding and stretching between galaxies, there is a rarefied plasma[42] that is thought to possess a cosmic filamentary structure[43] and that is slightly denser than the average density in the universe. This material is called the *intergalactic medium (IGM)* and is mostly ionized hydrogen; i.e. a plasma consisting of equal numbers of electrons and protons. The IGM is thought to exist at a density of 10 to 100 times the average density of the universe (10 to 100 hydrogen atoms per cubic meter). It reaches densities as high as 1000 times the average density of the universe in rich clusters of galaxies.

The reason the IGM is thought to be mostly ionized gas is that its temperature is thought to be quite high by terrestrial standards (though some parts of it are only "warm" by astrophysical standards). As gas falls into the Intergalactic Medium from the voids, it heats up to temperatures of 10^5 K to 10^7 K, which is high enough for the bound electrons to escape from the hydrogen nuclei upon collisions. At these temperatures, it is called the Warm-Hot Intergalactic Medium (WHIM). Computer simulations indicate that on the order of half the atomic matter in the universe might exist in this warm-hot, rarefied state. When gas falls from the filamentary structures of the WHIM into the galaxy clusters at the intersections of the cosmic filaments, it can heat up even more, reaching temperatures of 10^8 K and above.

See also

- Earth's location in the universe
- Interplanetary Internet
- Kármán line
- NASA
- Outer Space Treaty
- Private spaceflight
- Solar wind
- Space and survival
- Space colonization
- Space exploration
- Space science
- Space station
- Space technology
- Spaceflight
- Timeline of spaceflight
- Human outpost
- Timeline of Solar System exploration

External links

- Intergalactic Space [44], Natural History, Feb 1998
- Profits set to soar in outer space [45]
- Newscientist Space [46]
- X PRIZE Foundation [47]

References

[1] Nick Greene, about.com, Outer Space (http://space.about.com/od/glossaries/g/outerspace.htm)
[2] Porter, Roy; Park, Katharine; Daston, Lorraine (2006). *The Cambridge History of Science: Early modern science*. **3**. Cambridge University Press. p. 27. ISBN 0521572444.
[3] Eckert, Michael (2006). *The dawn of fluid dynamics: a discipline between science and technology*. Wiley-VCH. p. 5. ISBN 3527405135.
[4] Cajori, Florian (1917). *A history of physics in its elementary branches: including the evolution of physical laboratories*. New York: The Macmillan Company. pp. 64–66.
[5] Genz, Henning (2001). *Nothingness: the science of empty space*. Da Capo Press. pp. 127–128. ISBN 0738206105.
[6] Gatti, Hilary (2002). *Giordano Bruno and Renaissance science*. Cornell University Press. pp. 99–104. ISBN 0801487854.
[7] Kelly, Suzanne (1965). *The de muno of William Gilbert*. Amsterdam: Menno Hertzberger & Co.. p. 97–107.
[8] Olenick, Richard P.; Apostol, Tom M.; Goodstein, David L. (1986). *Beyond the mechanical universe: from electricity to modern physics*. Cambridge University Press. p. 356. ISBN 052130430X.
[9] Olenick *et al.* (1986:357–365).
[10] Thagard, Paul (1992). *Conceptual revolutions*. Princeton University Press. pp. 206–209. ISBN 0691024901.
[11] Maor, Eli (1991). *To infinity and beyond: a cultural history of the infinite*. Princeton paperbacks. p. 195. ISBN 0691025118.
[12] Webb, Stephen (1999). *Measuring the universe: the cosmological distance ladder*. Springer. pp. 71–73. ISBN 1852331062,.
[13] Tyson, Neil deGrasse; Goldsmith, Donald (2004). *Origins: fourteen billion years of cosmic evolution*. W. W. Norton & Company. pp. 114–115. ISBN 0393059928.
[14] Silk, Joseph (2000). *The Big Bang* (3rd ed.). Macmillan. ISBN 080507256X.
[15] OED[outer space]
[16] "Etymonline : Outer" (http://www.etymonline.com/index.php?search=outer+space&searchmode=none). . Retrieved 2008-03-24.
[17] Douglas Harper (November 2001). "Space" (http://www.etymonline.com/index.php?term=space). The Online Etymology Dictionary. . Retrieved 2009-06-19.
[18] Tadokoro, M. (1968). "A Study of the Local Group by Use of the Virial Theorem". *Publications of the Astronomical Society of Japan* **20**: 230. Bibcode: 1968PASJ...20..230T. This source estimates a density of 7×10^{-29} g/cm for the Local Group. An atomic mass unit is 1.66×10^{-24} g, for roughly 40 atoms per cubic metre.

[19] Borowitz, Sidney; Beiser, Arthur (1971). *Essentials of physics: a text for students of science and engineering*. Addison-Wesley series in physics (2nd ed.). Addison-Wesley Publishing Company. Note: this source gives a value of 2.7×10^{25} molecules per cubic meter.

[20] Davies, P. C. W. (1977). *The physics of time asymmetry*. University of California Press. p. 93. ISBN 0520032470. Note: a light year is about 10^{13} km.

[21] Chapmann, Glenn (May 22–27, 1991). "Space: the Ideal Place to Manufacture Microchips" (http://deneb.ensc.sfu.ca/papers/sdc-91w.pdf). in R. Blackledge, C. Radfield and S. Seida. . San Antonio, Texas. pp. 25–33. . Retrieved 2010-01-12.

[22] "Human Body in a Vacuum" (http://imagine.gsfc.nasa.gov/docs/ask_astro/answers/970603.html). NASA. June 03, 1997. . Retrieved 2009-06-19.

[23] Harding, Richard M (1989). *Survival in Space: Medical Problems of Manned Spaceflight*. Routledge. ISBN 0-415-00253-2.

[24] Billings, Charles E. (1973). "Barometric Pressure". in edited by James F. Parker and Vita R. West. *Bioastronautics Data Book* (2nd ed.). NASA. NASA SP-3006.

[25] Landis, Geoffrey A. (August 7, 2007). "Human Exposure to Vacuum" (http://www.geoffreylandis.com/vacuum.html). www.geoffreylandis.com. . Retrieved 2009-06-19.

[26] Webb, P. (1968). "The Space Activity Suit: An Elastic Leotard for Extravehicular Activity". *Aerospace Medicine* **39**: 376–383.

[27] Tamarack R. Czarnik. "Ebullism at 1 Million Feet: Surviving Rapid/Explosive Decompression" (http://www.geoffreylandis.com/ebullism.html). www.geoffreylandis.com. . Retrieved 2009-06-19.

[28] Andrea Thompson (9 April 2009). "Edge of Space Found" (http://www.space.com/scienceastronomy/090409-edge-space.html). space.com. . Retrieved 2009-06-19.

[29] L. Sangalli, D. J. Knudsen, M. F. Larsen, T. Zhan, R. F. Pfaff, and D. Rowland (2009). "Rocket-based measurements of ion velocity, neutral wind, and electric field in the collisional transition region of the auroral ionosphere" (http://www.agu.org/pubs/crossref/2009/2008JA013757.shtml). *Journal of Geophysical Research* (American Geophysical Union) **114**: A04306. doi:10.1029/2008JA013757. . Retrieved 2009-06-19.

[30] "Treaty on Principles Governing the Activities of States in the Exploration and Use of Outer Space, including the Moon and Other Celestial Bodies" (http://www.unoosa.org/oosa/SpaceLaw/outerspt.html). United Nations Office for Outer Space Affairs. January 1, 2008. . Retrieved 2009-12-30.

[31] "Index of Online General Assembly Resolutions Relating to Outer Space" (http://www.unoosa.org/oosa/en/SpaceLaw/gares). United Nations Office for Outer Space Affairs. . Retrieved 2009-12-30.

[32] "Columbus launch puts space law to the test" (http://www.esf.org/research-areas/humanities/news/ext-news-singleview/article/columbus-launch-puts-space-law-to-the-test-358.html). European Science Foundation. November 5, 2007. . Retrieved 2009-12-30.

[33] Linda Shiner (November 1, 2007). "X-15 Walkaround" (http://www.airspacemag.com/history-of-flight/x-15_walkaround.html). Air & Space Magazin. . Retrieved 2009-06-19.

[34] "Report of the Living With a Star Geospace Mission Definition Team" (http://www.lws.nasa.gov/documents/geospace/geospace_gmdt_report.pdf) (PDF). NASA. September 2002. . Retrieved 2007-12-19.

[35] "LWS Geospace Missions" (http://www.lws.nasa.gov/missions/geospace/geospace.htm). NASA. . Retrieved 2007-12-19.

[36] Papagiannis, Michael D. (1972). *Space Physics and Space Astronomy*. Taylor & Francis. pp. 12–149. ISBN 0677040008.

[37] Phillips, Tony (2009-09-29). "Cosmic Rays Hit Space Age High" (http://science.nasa.gov/headlines/y2009/29sep_cosmicrays.htm). NASA. . Retrieved 2009-10-20.

[38] Frisch, Priscilla C.; Müller, Hans R.; Zank, Gary P.; Lopate, C. (May 6–9, 2002). "Galactic environment of the Sun and stars: interstellar and interplanetary material". in Mario Livio, I. Neill Reid, William B. Sparks. *Astrophysics of life. Proceedings of the Space Telescope Science Institute Symposium*. Space Telescope Science Institute symposium series, Volume 16. Baltimore, MD, USA: Cambridge University Press. pp. 21–34. ISBN 0-521-82490-7. Bibcode: 2005asli.symp...21F.

[39] Flynn, G. J.; Keller, L. P.; Jacobsen, C.; Wirick, S. (2003). "The Origin of Organic Matter in the Solar System: Evidence from the Interplanetary Dust Particles". in R. Norris and F. Stootman. *Bioastronomy 2002: Life Among the Stars, Proceedings of IAU Symposium #213*. San Francisco: Astronomical Society of the Pacific. Bibcode: 2004IAUS..213..275F.

[40] Davidson, Keay & Smoot, George. Wrinkles in Time. New York: Avon, 2008: 158-163

[41] Silk, Joseph. Big Bang. New York: Freeman, 1977: 299.

[42] Jafelice, Luiz C. and Opher, Reuven (July 1992). "The origin of intergalactic magnetic fields due to extragalactic jets" (http://adsabs.harvard.edu/cgi-bin/nph-bib_query?bibcode=1992MNRAS.257..135J). Royal Astronomical Society. . Retrieved 2009-06-19.

[43] James Wadsley et al. (20 August 2002). "The Universe in Hot Gas" (http://antwrp.gsfc.nasa.gov/apod/ap020820.html). NASA. . Retrieved 2009-06-19.

[44] http://articles.findarticles.com/p/articles/mi_m1134/is_n1_v107/ai_20517887

[45] http://money.cnn.com/2006/02/27/technology/business2_guidetospaceintro

[46] http://www.newscientistspace.com

[47] http://www.xprize.org

Human spaceflight

Human spaceflight is spaceflight with a human crew and possibly passengers. This makes it unlike robotic space probes or remotely-controlled satellites. Human spaceflight is sometimes called **manned spaceflight**, a term now deprecated by major space agencies in favor of its gender-neutral alternative.

Edward White on a spacewalk during the Gemini 4 mission

The first human spaceflight was performed on April 12, 1961 by Soviet cosmonaut Yuri Gagarin. Today, the only three countries to have independent human spaceflight capability are Russia, United States and China. As of 2010, human spaceflights are being actively launched by the Soyuz programme conducted by the Russian Federal Space Agency, the Space Shuttle program conducted by NASA, and the Shenzhou program conducted by the China National Space Administration.

The U.S. has featured many of its space milestones on postage stamps.

The US will lose governmental human spaceflight capability upon retirement of the Space Shuttle, expected in late 2010. Under the Bush administration, the Constellation program included plans for canceling the Shuttle and replacing it with the capability for spaceflight beyond low Earth orbit. In the 2011 United States federal budget, the Obama administration proposed canceling Constellation. Under the new plans NASA would rely on transportation services provided by the private sector, such as Space X's Falcon 9. The period between the retirement of the Shuttle and the initial operational capability of new systems (either Constellation or the new commercial proposals), similar to the gap between the cancellation of Apollo and the first Space Shuttle flight, is often referred to as the human spaceflight gap.

In recent years there has been a gradual movement towards more commercial forms of spaceflight. A number of non-governmental startup companies have sprung up in recent years, hoping to create a space tourism industry. For a list of such companies, and the spacecraft they are currently building, see list of space tourism companies. NASA has also tried to stimulate private spaceflight through programs such as Commercial Crew Development (CCDev) and Commercial Orbital Transportation Services (COTS). With its 2011 budget proposals released in early February 2010[1] , the Obama administration is moving towards a model where commercial companies would supply NASA with transportation services of both crew and cargo to low Earth orbit. The vehicles used for these services would then serve both NASA and potential commercial customers such as Bigelow Aerospace. Similarly, NASA has made inquiries about leasing or buying space station modules from Bigelow Aerospace. NASA intends to spend $6 billion in the coming years to develop commercial crew vehicles, using a model similar to that used under COTS.[2]

History

First human spaceflights

Yuri Gagarin, the first man in space, in his space suit during the Vostok 1 mission

The first human spaceflight was undertaken on April 12, 1961, when cosmonaut Yuri Gagarin made one orbit around the Earth aboard the Vostok 1 spacecraft, launched by the Soviet space program and designed by the rocket scientist Sergey Korolyov. Valentina Tereshkova became the first woman in space on board Vostok 6 on June 16, 1963. Both spacecraft were launched by Vostok 3KA launch vehicles. Alexei Leonov made the first spacewalk when he left the Voskhod 2 on March 8, 1965. Svetlana Savitskaya became the first woman to do so on July 25, 1984.

The United States became the second nation (and for four decades, one of only two) to achieve manned spaceflight, with the suborbital flight of astronaut Alan Shepard aboard *Freedom 7*, carried out as part of Project Mercury. The spacecraft was launched on May 5, 1961 on a Redstone rocket. The first U.S. orbital flight was that of John Glenn aboard *Friendship 7*, which was launched February 20, 1962 on an Atlas rocket. Since April 12, 1981 the U.S. has conducted all its human spaceflight missions with reusable Space Shuttles. Sally Ride became the first American woman in space in 1983. Eileen Collins was the first female Shuttle pilot, and with Shuttle mission STS-93 in July 1999 she became the first woman to command a U.S. spacecraft.

Buzz Aldrin on the surface of the Moon during Apollo 11

The People's Republic of China became the third nation to achieve human spaceflight when Yang Liwei launched into space on a Chinese-made vehicle, the Shenzhou 5, on October 15, 2003. The flight made China the third nation to have launched its own manned spacecraft using its own launcher. Previous European (Hermes) and Japanese (HOPE-X) domestic manned programs were abandoned after years of development, as was the first Chinese attempt, the Shuguang spacecraft.

The farthest destination for a human spaceflight mission has been the Moon, and as of 2010 the only missions to the Moon have been those conducted by NASA as part of the Apollo program. The first such mission, Apollo 8, orbited the Moon but did not land. The first Moon landing mission was Apollo 11, during which—on July 20, 1969—Neil Armstrong and Buzz Aldrin became the first people to set foot on the Moon. Six missions landed in total, numbered Apollo 11–17, excluding Apollo 13. Altogether twelve men walked on the Moon, the only humans to have been on an extraterrestrial body. The Soviet Union discontinued its program for lunar orbiting and landing of human spaceflight missions on June 24, 1974 when Valentin Glushko became General Designer of NPO Energiya.[3]

The longest single human spaceflight is that of Valeriy Polyakov, who left earth on January 8, 1994, and didn't return until March 22, 1995 (a total of 437 days 17 hr. 58 min. 16 sec. aboard). Sergei Krikalyov has spent the most time of anyone in space, 803 days, 9 hours, and 39 seconds altogether. The longest period of continuous human presence in space lasted as long as 3,644 days, eight days short of 10 years, spanning the launch of Soyuz TM-8 on September 5, 1989 to the landing of Soyuz TM-29 on August 28, 1999.

For many years beginning in 1961, only two countries, the USSR (later Russia) and United States, had their own astronauts. Later, cosmonauts and astronauts from other nations flew in space, beginning with the flight of Vladimir Remek, a Czech, on a Soviet spacecraft on March 2, 1978. As of 2007, citizens from 33 nations (including space tourists) have flown in space aboard Soviet, American, Russian, and Chinese spacecraft.

Space programs

As of 2010, human spaceflight missions have been conducted by the former Soviet Union/(Russia), the United States, the People's Republic of China and by the private spaceflight company Scaled Composites.

Several other countries and space agencies have announced and begun human spaceflight programs by their own technology, including India (ISRO), Ecuador (EXA), Japan (JAXA), Iran (ISA) and Malaysia (MNSA.

Currently the following spacecraft and spaceports are used for launching human spaceflights:

- Soyuz with Soyuz rocket—Baikonur Cosmodrome
- Space Shuttle—Kennedy Space Center
- International Space Station (ISS)—Assembled in orbit; crews transported by the previous two spacecraft
- Shenzhou spacecraft with Long March rocket—Jiuquan Satellite Launch Center

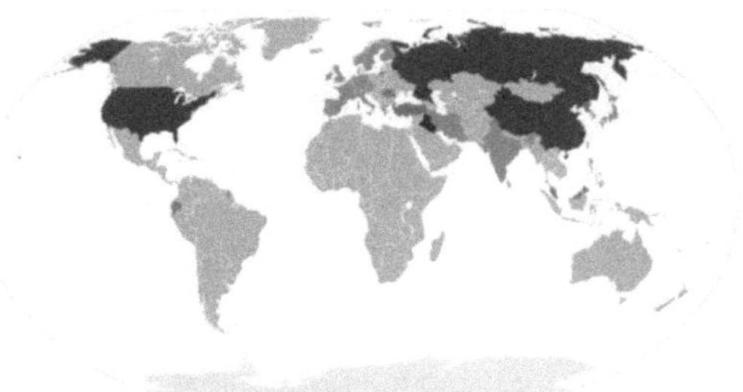

Countries which have human spaceflight agendas.

Historically, the following spacecraft and spaceports have also been used for human spaceflight launches:

- Vostok—Baikonur Cosmodrome
- Mercury—Cape Canaveral Air Force Station
- Voskhod—Baikonur Cosmodrome
- X-15—Edwards Air Force Base,[4] (two internationally recognized suborbital flights in program)
- Gemini—Cape Canaveral Air Force Station
- Apollo—Kennedy Space Center (Apollo 7 at Cape Canaveral Air Force Station)
- Salyut space station—Baikonur Cosmodrome
- Almaz space station—Baikonur Cosmodrome (Almaz was a series of military space stations under cover of the civilian name Salyut)
- Skylab space station—Kennedy Space Center
- Mir space station—Baikonur Cosmodrome
- SpaceShipOne with White Knight—Mojave Spaceport

Numerous private companies attempted human spaceflight programs in an effort to win the $10 million Ansari X Prize. The first private human spaceflight took place on June 21, 2004, when SpaceShipOne conducted a suborbital flight. SpaceShipOne captured the prize on October 4, 2004, when it accomplished two consecutive flights within one week.

Most of the time, the only humans in space are those aboard the ISS, whose crew of six spends up to six months at a time in low Earth orbit.

NASA and ESA now use the term "human spaceflight" to refer to their programs of launching people into space. Traditionally, these endeavors have been referred to as "manned space missions."

National spacefaring attempts

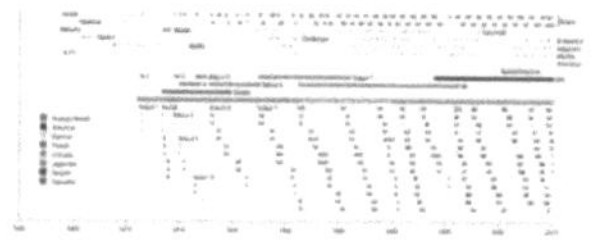

Successfully executed manned programs are in **bold**.

Suborbital spaceflights are in *italics*.

Nation/Organization	Space agency	National term	First launched astronaut	Date	Spacecraft	Launcher
Soviet Union	**Soviet space program (OKB-1 Design Bureau)**	**cosmonaut** **космонавт** (Russian) ***kosmonavt***	**Yuri Gagarin**	**April 12, 1961**	**Vostok 1**	**Vostok**
United States	***National Aeronautics and Space Administration (NASA)***	***astronaut***	***Alan Shepard***	***May 5, 1961***	***Mercury-Redstone 3***	***Redstone***
China	China space program	宇航员 (Chinese) *yǔhángyuán* 航天员 (Chinese) *hángtiānyuán*	...	1973 (abandoned)	Shuguang 1	Long March 2A
China	China space program	宇航员 (Chinese) *yǔhángyuán* 航天员 (Chinese) *hángtiānyuán*	...	1981 (abandoned)	Piloted FSW	Long March 2
ESA	European Space Agency (ESA)	astronaut spationaut *spationaute* (French)	...	1992 (abandoned)	Hermes	Ariane V
Iraq[5]	...	رجل فضاء (Arabic) *rajul faḍā'* رائد فضاء (Arabic) *rā'ib faḍā'* ملاح فضائي (Arabic) *mallāḥ faḍā'iy*	...	2001 (abandoned)	...	Tammouz 2 or 3
Japan	Japan Aerospace Exploration Agency (JAXA)	宇宙飛行士 (Japanese) *uchūhikōshi*	...	2003 (abandoned)	HOPE-X	H-II
China	**China National Space Administration (CNSA)**	**taikonaut** 太空人 (Chinese) ***tàikōng rén*** 宇航员 (Chinese) ***yǔhángyuán*** 航天员 (Chinese) ***hángtiānyuán***	**Yang Liwei**	**October 15, 2003**	**Shenzhou 5**	**Long March 2F**

India	Indian Space Research Organisation (ISRO)	vyomanaut gaganaut *aakashagaami* आकाशगामिः (Sanskrit) *brahmāndagaami* ब्रह्मान्डगामिः (Sanskrit) *antarikshyaatri* अन्तरिक्षयात्रिः (Sanskrit)	...	2016 (approved)[6]	Orbital Vehicle (OV)	GSLV Mk II
ESA	European Space Agency (ESA)	astronaut spationaut *spationaute* (French)	...	2020 (approved conceptually but full development not begun)[7] [8] [9] [10]	ARV phase-2 (may be changed to CSTS)	Ariane V
Iran	Iranian Space Agency (ISA)	دروناضف (Persian) *faza navard*	...	2021 (planned)	ISA manned spacecraft	...
Japan	Japan Aerospace Exploration Agency (JAXA)	宇宙飛行士 (Japanese) *uchūhikōshi*	...	2025 (planned)	HTV-based spacecraft	H-IIB
Romania	*Romanian Cosmonautics and Aeronautics Association (ARCASPACE)*	astronaut *astronaut* (Romanian)	...	*TBA (approved)*	*Stabilo-mission8*	*ARCASPACE air-balloon*

Safety concerns

Planners of human spaceflight missions face a number of safety concerns.

Life support

The immediate needs for breathable air and drinkable water are addressed by the life support system of the spacecraft.

Medical issues

Effects of microgravity

Medical data from astronauts in low earth orbits for long periods, dating back to the 1970s, show several adverse effects of a microgravity environment: loss of bone density, decreased muscle strength and endurance, postural instability, and reductions in aerobic capacity. Over time these deconditioning effects can impair astronauts' performance or increase their risk of injury.[11]

In a weightless environment, astronauts put almost no weight on the back muscles or leg muscles used for standing up. Those muscles then start to weaken and eventually get smaller. If there is an emergency at landing, the loss of muscles, and consequently the loss of strength can be a serious problem. Sometimes, astronauts can lose up to 25% of their muscle mass on long term flights. When they get back to ground, they will be considerably weakened and will be out of action for a while.

Astronauts experiencing weightlessness will often lose their orientation, get motion sickness, and lose their sense of direction as their bodies try to get used to a weightless environment. When they get back to Earth, or any other mass with gravity, they have to readjust to the gravity and may have problems standing up, focusing their gaze, walking and turning. Importantly, those body motor disturbances after changing from different gravities only get worse the longer the exposure to little gravity. These changes will affect operational activities including approach and landing, docking, remote manipulation, and emergencies that happen by landing. This is a big problem for mission success.

Radiation

Without proper shielding the crews of missions beyond low Earth orbit (LEO) might be at risk from high-energy protons emitted by solar flares. Lawrence Townsend of the University of Tennessee and others have studied the most powerful solar flare ever recorded. That flare was seen by the British astronomer Richard Carrington in September 1859. Radiation doses astronauts would receive from a Carrington-type flare could cause acute radiation sickness and possibly even death.[12]

Another type of radiation, galactic cosmic rays, present further challenges to human spaceflight beyond LEO.[13]

Radiation damage to the immune system

Another factor is that extended space flight might slow down the body's ability to protect itself against diseases. Some of the problems are a weakened immune system and the activation of dormant viruses in the body. Radiation can cause both short and long term consequences to the bone marrow stem cells which create the blood and immune systems. Because the interior of a spacecraft is so small, a weakened immune system and more active viruses in the body can lead to a fast spread of infection.

Isolation

During long missions, astronauts are isolated and confined into small spaces. Depression, cabin fever and other psychological problems may result that impact crew safety and mission success.

Astronauts may not be able to quickly return to Earth or receive medical supplies, equipment or personnel if a medical emergency occurs. The astronauts may have to rely for long periods on their limited existing resources and medical advice from the ground.

Fatality risk

As of 2009, 18 crew members have died during actual spaceflight missions (see table). Over 100 others have died in accidents during activity directly related to spaceflight missions or testing.

Year	# of Deaths	Mission	**Known or likely cause**
1967	1	Soyuz 1	Trauma from Earth surface impact
1971	3	Soyuz 11	Asphyxia from cabin breach
1986	7	Space Shuttle Challenger	Trauma from Earth surface impact (mission never reached space)
2003	7	Space Shuttle Columbia	Asphyxia from cabin breach or trauma from object impact

See also

- Astronaut
- Astronautical hygiene
- Atmospheric reentry
- Human adaptation to space
- Interplanetary travel
- Mars to Stay
- Monkeys in space
- Space and survival
- Space colonization
- Space exploration
- Space tourism
- Spaceflight records
- Spaceport
- Space Suit
- SpaceShipOne
- Timeline of Solar System exploration
- Timeline of space travel by nationality
- Timeline of first orbital launches by country
- X-15 program

Lists

- List of astronauts by name
- List of human spaceflights
- List of human spaceflight programs
- List of manned spacecraft
- List of space agencies
- List of space disasters
- List of spacewalks

External links

- NASA Human Space Flight [14]
- Official Constellation NASA Web Site [15]
- Official Orion NASA Web Site [16]
- Official Ares NASA Web Site [17]
- Human Spaceflight Profile [18] by NASA's Solar System Exploration [19]
- Transitioning to the NASA Constellation Program [20]
- U.S. Spaceflight History [21]

References

[1] "FY 2011 Budget" (http://www.nasa.gov/news/budget/index.html). .

[2] "Mission Directorate: Exploration Systems" (http://www.nasa.gov/pdf/428356main_Exploration.pdf) (PDF). .

[3] Siddiqi, Asif. *Challenge To Apollo The Soviet Union and The Space Race, 1945-1974* (http://ntrs.nasa.gov/search.jsp?Ntk=all&Ntx=mode matchall&Ntt=SP-2000-4408). NASA. pp. 832. .

[4] "X-15 Hypersonic Research Program" (http://www.nasa.gov/centers/dryden/news/FactSheets/FS-052-DFRC.html). NASA. .

[5] According to a press-release of Iraqi News Agency of December 5, 1989 about the first (and last) test of the Tammouz space launcher, Iraq intended to develop manned space facilities by the end of the century. These plans were put to an end by the Gulf War of 1991 and the economic hard times that followed.

[6] Beary, Habib (2010-01-27). "India announces first manned space mission" (http://news.bbc.co.uk/2/hi/south_asia/8483787.stm). BBC. pp. 1. .

[7] "Europe targets manned spaceship" (http://news.bbc.co.uk/2/hi/science/nature/8139347.stm). *BBC News*. July 7, 2009. . Retrieved March 27, 2010.

[8] http://www.flightglobal.com/articles/2008/05/22/223941/apollo-like-capsule-chosen-for-crew-space-transportation.html

[9] http://esamultimedia.esa.int/docs/ATV/infokit/english/Complete_Infokit_ATVreentry.pdf

[10] "Europe's 10bn-euro space vision" (http://news.bbc.co.uk/1/hi/sci/tech/7749761.stm). *BBC News*. November 26, 2008. . Retrieved March 27, 2010.

[11] "Exploration Systems Human Research Program - Exercise Countermeasures" (http://exploration.grc.nasa.gov/Exploration/Advanced/Human/Exercise/). NASA. .

[12] Stephen Battersby (21 March 2005). "Superflares could kill unprotected astronauts" (http://www.newscientist.com/article/dn7142). .

[13] "Space Radiation Hazards and the Vision for Space Exploration" (http://www.nap.edu/catalog.php?record_id=11760). NAP. 2006. .

[14] http://spaceflight.nasa.gov/

[15] http://www.nasa.gov/constellation

[16] http://www.nasa.gov/orion

[17] http://www.nasa.gov/ares

[18] http://solarsystem.nasa.gov/missions/profile.cfm?MCode=Human_Spaceflight

[19] http://solarsystem.nasa.gov

[20] http://www.nasaconstellation.com

[21] http://www.spaceflighthistory.com

Human adaptation to spaceflight

Human physiological adaptation to the conditions of space is a challenge faced in the development of human spaceflight.

The fundamental engineering problems of escaping Earth's gravity well and developing systems for in-space propulsion have been examined for well over a century, and millions of man-hours of research have been spent on them. In recent years there has been an increase in research into the issue of how humans can survive and work in space for extended and possibly indefinite periods of time. This question requires input from the whole gamut of physical and biological sciences and has now become the greatest challenge, other than funding, to human space exploration. A fundamental step in overcoming this challenge is trying to understand the effects and the impact long-term space travel has on the human body.

Importance

Space colonization efforts must take into account the effects of space on the body

The sum of mankind's experience has resulted in the accumulation of 58 solar years in space and a much better understanding of how the human body adapts. In the future, industrialisation of space and exploration of inner and outer planets will require humans to endure longer and longer periods in space. A recent report has highlighted the limits to the quality and quantity of current data, which make it difficult for scientists to extrapolate all the effects of living in space long term. The majority of the data comes from missions of short duration and so some of the long-term physiological effects of living in space are still unknown. A round trip to Mars with current technology is estimated to involve at least 18 months in transit alone. How the human body reacts to such time periods in space is a vital part of the preparation for such journeys. On-board medical facilities need to be able to cope with any type of trauma or emergency as well as contain a huge variety of diagnostic and medical instruments in order to keep a crew healthy over a long period of time, as these will be the only facilities available on board a spacecraft to cope with not only trauma, but also the adaptive responses of the human body in space.

Public perception

Public perception of living in space is generally skewed towards the TV sci-fi of *Star Trek* where gravity is artificially produced and journeys to distant stars take less than a day.[1] The real-world difficulties of extended space travel are far more complex than those presented in most science fiction novels or movies.

Effects on humans

The effects of space conditions on humans can be separated into two areas, the physical and the psychological.

Unprotected effects

The environment of space is highly dangerous without appropriate protection. The greatest threat is from the lack of pressure in the vacuum environment, while temperature and radiation effects also have an influence. In the low pressure environment, gas exchange in the lungs would continue as normal but would result in the removal of all gases, including oxygen, from the bloodstream. After 9 to 12 seconds, the deoxygenated blood would reach the brain, and loss of consciousness would result.[2] Death would gradually follow after two minutes of exposure—though the limits are uncertain. As shown in the film depiction of Arthur C. Clarke's vignette in *2001: A Space Odyssey* (1968), if actions are taken quickly, and normal pressure restored within around 90 seconds, the victim may well make a full recovery.[3]

Humans and animals exposed to vacuum will lose consciousness after a few seconds and die of hypoxia within minutes, but the symptoms are not nearly as graphic as the imagery in the public media. Blood and other body fluids do boil when their pressure drops below 6.3 kPa (47 Torr), the vapour pressure of water at body temperature.[4] This condition is called ebullism.[5] The steam may bloat the body to twice its normal size and slow circulation, but tissues are elastic and porous enough to prevent rupture. Ebullism is slowed by the pressure containment of blood vessels, so some blood remains liquid.[6] [2] Swelling and ebullism can be reduced by containment in a flight suit. Shuttle astronauts wear a fitted elastic garment called the Crew Altitude Protection Suit (CAPS) which prevents ebullism at pressures as low as 2 kPa (15 Torr).[7] Rapid evaporative cooling of the skin will create frost, particularly in the mouth, but this is not a significant hazard.

This painting, An Experiment on a Bird in the Air Pump by Joseph Wright of Derby, 1768, depicts an experiment performed by Robert Boyle in 1660.

A short term exposure to vacuum of up to 30 seconds is unlikely to cause permanent physical damage.[8] Animal experiments show that rapid and complete recovery is the norm for exposures shorter than 90 seconds, while longer full-body exposures are fatal and resuscitation has never been successful.[9] There is only a limited amount of data available from human accidents, but it is consistent with animal data. Limbs may be exposed for much longer if breathing is not impaired.[4] Robert Boyle was the first to show in 1660 that vacuum is lethal to small animals. In 1942, in one of a series of experiments on human subjects for the Luftwaffe, the Nazi regime exposed Dachau concentration camp prisoners to vacuum in order to determine the human body's capacity to survive high-altitude conditions.

As well as experimentation with humans and monkeys, a few cases of loss of pressure have occurred in the past, especially in experimentation on spaceflight projects. One such case is discussed in a NASA technical report: *Rapid (Explosive) Decompression Emergencies in Pressure-Suited Subjects*:

> "At NASA's Manned Spacecraft Center (now renamed Johnson Space Center) we had a test subject accidentally exposed to a near vacuum (less than 1 psi) [7 kPa] in an incident involving a leaking space suit in a vacuum chamber back in '65. He remained conscious for about 14 seconds, which is about the time it takes for O_2 deprived blood to go from the lungs to the brain. The suit probably did not reach a hard vacuum, and we began repressurizing the chamber within 15 seconds. The subject regained consciousness at around 15,000 feet [4600 m] equivalent altitude. The subject later reported that he could feel and hear the air leaking out, and his last conscious memory was of the water on his tongue beginning to boil."

There has been one recorded incident of death from decompression in spaceflight, the Soyuz 11 decompression accident, in 1971.

Cold or oxygen-rich atmospheres can sustain life at pressures much lower than atmospheric, as long as the density of oxygen is similar to that of standard sea-level atmosphere. The colder air temperatures found at altitudes of up to 3 km generally compensate for the lower pressures there.[4] Above this altitude, oxygen enrichment is necessary to prevent altitude sickness, and spacesuits are necessary to prevent ebullism above 19 km.[4] Most spacesuits use only 20 kPa (150 Torr) of pure oxygen, just enough to sustain full consciousness. This pressure is high enough to prevent ebullism, but simple evaporation of blood, or of gases dissolved in the blood, can still cause decompression sickness (the bends) and gas embolisms if not managed.

Rapid decompression can be much more dangerous than vacuum exposure itself. Even if the victim does not hold his breath, venting through the windpipe may be too slow to prevent the fatal rupture of the delicate alveoli of the lungs.[4] Eardrums and sinuses may be ruptured by rapid decompression, soft tissues may bruise and seep blood, and the stress of shock will accelerate oxygen consumption leading to hypoxia.[5] Injuries caused by rapid decompression are called barotrauma, and are well known from scuba diving accidents. A pressure drop as small as 100 Torr (13 kPa), which produces no symptoms if it is gradual, may be fatal if it occurs suddenly.[4]

In a vacuum there is no medium for removing heat from the body by conduction or convection. Loss of heat is by radiation from the 310 K person to the 3 K of outer space. This is a slow process, especially in a clothed person, so there is no danger of immediately freezing.[10] (Evaporation of skin moisture in the vacuum would cause immediate cooling but only by a very small amount.) Exposure to the 600 K radiation from the Sun would lead to local heating that would be well distributed by the body's conductivity and blood circulation. Other solar radiation, particularly ultraviolet rays, may cause severe sunburn in a few seconds.

Protected effects

Despite modern technology, some hazards still prove impossible to remove. The most important factor affecting human physical well-being in space is weightlessness, more accurately defined as microgravity environments. Living in this type of environment impacts on three types of human tissue:

- gravity receptors
- fluids
- weight bearing structures

Gravity receptors

Living on earth we constantly feel the gravitational pull and our bodies react automatically to maintain posture and locomotion in a downward pulling world. In microgravity environments, these constant signals the body is adapted to are absent. The otolith organs in the middle ear sensitive to linear accelerations no longer perceive a downwards bias, muscles are no longer required to contract to maintain posture and pressure receptors in the feet and ankles no longer signal the direction of down. These changes can immediately result in visual-orientation illusions where the astronaut feels he has flipped 180 degrees. Over time however the brain adapts and although these illusions can still occur most astronauts begin to see "down" as where the feet are. People returning to Earth after extended weightless periods initially have great difficulty maintaining their balance but recover the ability very quickly, highlighting the remarkable ability of the human body to adapt. Over half of astronauts also experience symptoms of motion sickness for the first three days of travel due to the conflict between what the body expects and what the body actually perceives.[11]

Fluids

The second effect of weightlessness takes place in human fluids. The body is made up of 60% water, much of it intra-vascular and inter-cellular. Within a few moments of entering a microgravity environment, fluid is immediately re-distributed to the upper body resulting in bulging neck veins, puffy face and sinus and nasal congestion which can last throughout the duration of the trip and is very much like the symptoms of the common cold. In space the autonomic reactions of the body to maintain blood pressure are not required and fluid is distributed more widely around the whole body. This results in a decrease in plasma (water in the blood stream) volume of around 20%. These fluid shifts initiate a cascade of adaptive systemic effects that can be dangerous upon return to earth. Orthostatic intolerance results in astronauts returning to Earth after extended space missions being unable to stand unassisted for more than 10 minutes at a time without fainting. This is due in part to changes in the autonomic regulation of blood pressure and the loss of plasma volume. Although this effect becomes worse the longer the time spent in space, as yet all individuals have returned to normal within at most a few weeks of landing.

Weight bearing structures

The third and most worrying effect of long-term weightlessness involves bones and muscles. Without the effects of gravity, skeletal muscle is no longer required to maintain our posture and the muscle groups used in moving around in a weightless environment are very different to those required in terrestrial locomotion. Consequently some muscles atrophy rapidly. The types of muscle fibre prominent in muscles also change. Slow twitch endurance fibres used to maintain posture are replaced by fast twitch rapidly contracting fibres that are insufficient for any heavy labour. Bone metabolism also changes. Normally bone is laid down in the direction of mechanical stress, however in a microgravity environment there is very little mechanical stress. This results in a loss of bone tissue approximately 1.5% per month especially from the lower vertebrae, hip and femur. Elevated blood calcium levels from the lost bone result in dangerous calcification of soft tissues and potential kidney stone formation. It is still unknown whether bone recovers completely. Loss of bone and muscle make it very difficult for humans to move and even breathe under the weight of Earth's pull upon their return.

Effects of radiation

Weightlessness is not the only factor to affect the human body in space. Without the protection of the Earth's atmosphere and magnetosphere astronauts are exposed to high levels of radiation through a steady flux of cosmic rays which pose a serious health threat. A year in even low-earth orbit results in a dose of radiation 10 times that of the annual dose on earth resulting in a high risk of astronauts developing cancer. High levels of radiation can create 'chromosomal aberrations' in blood lymphocytes. These cells are heavily involved in the immune system and so any damage may contribute to the lowered immunity experienced by astronauts. Over time immunodeficiency results in the rapid spread of infection between crew members, especially in such confined areas. Radiation has also recently been linked to a higher incidence of cataracts in astronauts. Protective shielding and protective drugs may lower the risks to astronauts to an acceptable level, but data is scarce, and longer-term exposure will result in greater risks.

Sense of taste

One effect of weightlessness on humans is that some astronauts report a change in their sense of taste in space.[12] Some astronauts find that their food is bland, others find that their favorite foods no longer taste as good, some astronauts enjoy eating certain foods that they would not normally eat and some find no change whatsoever. The reason for this is uncertain, and several theories have been suggested:

- Congestion: Microgravity may cause fluid buildup in the sinuses, changing the taste in a similar fashion to holding one's nose while eating.
- Physical food degradation: Food in orbit is often stored for some months before being consumed. This and the stellar radiation may cause a breakdown in the groups of chemicals that give food its taste, resulting in bland food.
- Boredom: Menus for the ISS astronauts are planned on a repeating 8-day cycle, which are selected from a menu designed by NASA, and taken into space with the astronaut. This constant repetition may lead to some astronauts getting tired of food that they had previously liked.
- Psychological changes: The loss of taste may be purely psychological.

Astronauts often choose strongly tasting food such as salsa or shrimp cocktail.

Other physical effects

Other physical discomforts such as back and abdominal pain are commonly experienced with no clear cause. These may be part of the asthenia syndrome reported by cosmonauts living in space over an extended period of time, but seen as anecdotal by astronauts. Fatigue, listlessness, and psychosomatic worries are also part of the syndrome. The data is inconclusive, however the syndrome does appear to exist as a manifestation of all the internal and external stress crews in space must face. The amount and quality of sleep experienced in space is poor due to highly variable light and dark cycles on flight decks and poor illumination during daytime hours in the space craft. Even the habit of looking out of the window before retiring can send the wrong messages to the brain resulting in poor sleep patterns. These disturbances in circadian rhythm have profound effects on the neurobehavioural responses of crew and aggravate the psychological stresses they already experience.

Studies of Russian cosmonauts, such as those on Mir, provide data on the long-term effects of space on the human body.

Psychological effects

The psychological effects of living in space have not been clearly analyzed but analogies on Earth do exist, such as Arctic research stations and submarines. The enormous stress on the crew, coupled with the body adapting to other environmental changes, can result in anxiety, insomnia and depression. According to current data however astronauts and cosmonauts seem extremely resilient to psychological stresses. Interpersonal issues can have an enormous influence on a human's well-being and yet little research has been undertaken to examine crew selection issues in relation to this. The Mars Arctic Research Station and Mars Desert Research Station have examined the influence of different crew selections when living in a completely isolated environment and may provide vital data for future experiences.

Future prospects

At the moment only rigorously tested humans have experienced the conditions of space. If off-world colonization someday begins, many types of people will be exposed to these dangers, and the effects on the elderly and on the very young are completely unknown. Factors such as nutritional requirements and physical environments which have not been examined here will become important. Overall, there is little data on the manifold effects of living in space and this makes working to mitigate the risks during a lengthy space habitation difficult. Test beds such as the International Space Station (ISS) are presently being utilized to research some of these risks.

The environment of space is still largely unknown, and there will likely be hazards of which we are not currently aware. Meanwhile, future technologies such as artificial gravity and more complex bioregenerative life support systems may someday be capable of mitigating some hazards.

See also

- Overview effect
- Space colonization
- Space adaptation syndrome
- Effects of Space on the Body

Sources

1. Nasa Report: Space Travel 'Inherently Hazardous' to Human Health. Leonard David. 2001
2. *Space Physiology and Medicine*. Third edition. A. E. Nicogossian, C. L. Huntoon and S. L. Pool. Lea & Febiger, 1993.
3. L.-F. Zhang. Vascular adaptation to microgravity: What have we learned?. *Journal of Applied Physiology*. 91(6) (pp 2415–2430), 2001.
4. G. Carmeliet, Vico. L, Bouillon R. *Critical Reviews in Eukaryotic Gene Expression*. Vol 11(1-3) (pp 131–144), 2001.
5. F.A. Cucinotta et al. Space radiation cancer risks and uncertainties for Mars missions. *Radiation Research*. Vol 156:5 II;pp 682–688, 2001.
6. F.A. Cucinotta et al. Space radiation and cataracts in astronauts. *Radiation Research*. Vol 156(5 I) (pp 460–466), 2001.
7. Styf, Jorma R. MD; Hutchinson, Karen BS; Carlsson, Sven G. PhD, and; Hargens, Alan R. Ph.D. Depression, Mood State, and Back Pain During

References

[1] "The New Mythology of the Space Age", Sylvia Engdahl, http://www.sylviaengdahl.com/space/myth01.htm

[2] "Human Exposure to Vacuum" (http://www.geoffreylandis.com/vacuum.html). . Retrieved 2006-03-25.

[3] "What Happens to the Human Body in a Vacuum", About.com, http://space.about.com/cs/basics/a/bodyvacuum1.htm.

[4] Harding, Richard M. (1989), *Survival in Space: Medical Problems of Manned Spaceflight*, London: Routledge, ISBN 0-415-00253-2.

[5] Czarnik, Tamarack R. (1999). "Ebullism at 1 Million Feet: Surviving Rapid/Explosive Decompressionn" (http://www.geoffreylandis.com/ebullism.html). . Retrieved 2009-20-26.

[6] Billings, Charles E. (1973). "Barometric Pressure". in edited by James F. Parker and Vita R. West. *Bioastronautics Data Book* (Second Edition ed.). NASA. NASA SP-3006.

[7] Webb P. (1968). "The Space Activity Suit: An Elastic Leotard for Extravehicular Activity". *Aerospace Medicine* **39**: 376–383.

[8] "The Human Body in a Vacuum", NASA, http://imagine.gsfc.nasa.gov/docs/ask_astro/answers/970603.html

[9] Cooke JP, RW Bancroft (1966). "Some Cardiovascular Responses in Anesthetized Dogs During Repeated Decompressions to a Near-Vacuum". *Aerospace Medicine* **37**: 1148–1152.

[10] "Ask a scientist. Why is space cold?" (http://www.newton.dep.anl.gov/askasci/ast99/ast99383.htm). Argonne National Laboratory, Division of Educational Programs. . Retrieved 2008-11-27.

[11] "The Effects of Long-Duration Space Flight on Eye, Head, and Trunk Coordination During Locomotion (9307191)", NASA, http://lsda.jsc.nasa.gov/scripts/experiment/exper.cfm?exp_index=747

[12] "NASAeplores 5-8: A matter of taste", NASA, http://www.nasaexplores.com/show2_5_8a.php?id=03-038&gl=58

Space medicine

Space medicine is the practice of medicine on astronauts in outer space whereas astronautical hygiene is the application of science and technology to the prevention or control of exposure to the hazards that may cause astronaut ill health. Both these sciences work together to ensure that astronauts work in a safe environment. The main thing is to discover how well and for how long people can survive the extreme conditions in space, and how fast they can adapt to the earth's environment after coming back from space.

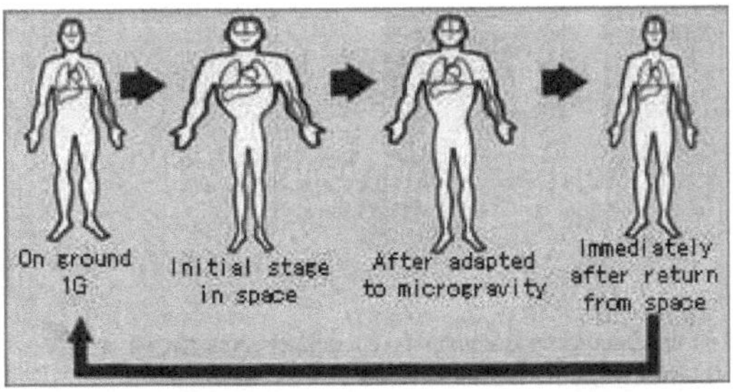

The effects of microgravity on fluid distribution around the body (greatly exaggerated).

History

Hubertus Strughold (1898-1987), a former Nazi physician and physiologist, was brought to the United States after World War II as part of Operation Paperclip[1] . He first coined the term "space medicine" in 1948 and was the first and only Professor of Space Medicine at the School of Aviation Medicine (SAM) at Randolph Air Force Base, Texas. In 1949 Strughold was made director of the Department of Space Medicine at the SAM (which is now the U.S. Air Force School of Aerospace Medicine [USAFSAM] at Brooks Air Force Base, Texas). He played an important role in developing the pressure suit worn by early American astronauts. He was a co-founder of the Space Medicine Branch of the Aerospace Medical Association in 1950. The aeromedical library at Brooks AFB was named after him in 1977, but later renamed because documents from the Nuremberg War Crimes Tribunal linked Strughold to medical experiments in which inmates of the Dachau concentration camp were tortured and killed[2] .

Who benefits from space medicine research?

Astronauts are not the only ones who benefit from space medicine research. Several medical products have been developed that are space spinoffs, that is practical applications for the field of medicine arising out of the space program. Because of joint research efforts between NASA, the National Institutes on Aging (a part of the National Institutes of Health), and other aging-related organizations, space exploration has benefitted a particular segment of society, seniors. Evidence of aging related medical research conducted in space was most publicly noticeable during STS-95 (See below).

Medical space spinoffs from the early space exploration years (pre-Mercury through Apollo projects)

- Radiation therapy for the treatment of cancer. In conjunction with Cleveland Clinic, the cyclotron at NASA's center in Cleveland, Ohio—which had been utilized for testing nuclear propulsion systems for air and space craft—was used in the first clinical trials for the treatment and evaluation of neutron radiation therapy for cancer patients.
- Foldable walkers. Made from a lightweight metal material developed by NASA for aircraft and spacecraft, foldable walkers are portable and easy to manage.
- Personal alert systems. These are emergency alert devices that can be worn by individuals who may require emergency medical or safety assistance. When a button is pushed, the device sends a signal to a remote location for help. To send the signal, the device relies on telemetry technology developed at NASA.
- CAT Scans and MRIs. These devices are used by hospitals to see inside the human body. Their development would not have been possible without the technology provided by NASA after it found a way to take better pictures of the Earth's moon.
- Muscle stimulator device. This device is used for ½ hour per day to prevent muscle atrophy in paralyzed individuals. It provides electrical stimulation to muscles which is equal to jogging three miles per week. Christopher Reeves used these in his therapy.
- Orthopedic evaluation tools. Equipment to evaluate posture, gait, and balance disturbances was developed at NASA, along with a radiation-free way to measure bone flexibility using vibration.
- Diabetic foot mapping. This technique was developed at NASA's center in Cleveland, Ohio to help monitor the effects of diabetes in feet. These efforts helped
- Foam cushioning. Special foam used for cushioning astronauts during liftoff is used in pillows and mattresses at many nursing homes and hospitals to help prevent ulcers, relieve pressure, and provide a better night's sleep.
- Kidney dialysis machines. These machines rely on technology developed by NASA in order to process and remove toxic waste from used dialysis fluid.
- Talking wheelchairs. Paralyzed individuals who have difficulty speaking may use a talking feature on their wheelchairs which was developed by NASA to create synthesized speech for aircraft.
- Collapsible, lightweight wheelchairs. These wheelchairs are designed for portability and can be folded and put into trunks of cars. They rely on synthetic materials that NASA developed for its air and space craft
- Surgically implantable heart pacemaker. These devices depend on technologies developed by NASA for use with satellites. They communicate information about the activity of the pacemaker, such as how much time remains before the batteries need to be replaced.
- Implantable heart defibrillator. This tool continuously monitors heart activity and can deliver an electric shock to restore heartbeat regularity.
- EMS Communications. Technology used to communicate telemetry between Earth and space was developed by NASA to monitor the health of astronauts in space from the ground. Ambulances use this same technology to send information—like EKG readings—from patients in transport to hospitals. This allows faster and better treatment.
- Weightlessness. While not an invention per se, the weightlessness of space one day may allow individuals with limited mobility on Earth—even those normally confined to wheelchairs—the freedom to move about with ease. A notable individual to take advantage of weightlessness in the "Vomit Comet" during 2007 was physicist Stephen Hawking.

Major historical medical investigations in space during the Space Shuttle era

STS-95

John Glenn, the first American astronaut to orbit the Earth, returned with much fanfare to space once again at 77 years of age to confront the physiological challenges preventing long-term space travel for astronauts—loss of bone density, loss of muscle mass, balance disorders, sleep disturbances, cardiovascular changes, and immune system depression—all of which are problems confronting aging people as well as astronauts. Once again Glenn stepped forward to play an historic role in the future of space exploration, but this time he would provide new medical research in the field of gerontology as well.

What are the effects of space on the body?

Decompression sickness

Decompression illness in spaceflight

In space, astronauts use a space suit, essentially a self-contained individual spacecraft, to do spacewalks, or extra-vehicular activities (EVAs). Spacesuits are generally inflated with 100% oxygen at a total pressure that is less than a third of normal atmospheric pressure. Eliminating inert atmospheric components such as nitrogen allows the astronaut to breathe comfortably, but also have the mobility to use their hands, arms, and legs to complete required work, which would be more difficult in a higher pressure suit.

After the astronaut dons the spacesuit, air is replaced by 100% oxygen in a process called a "nitrogen purge". In order to reduce the risk of decompression sickness, the astronaut must spend several hours "pre-breathing" at an intermediate nitrogen partial pressure, in order to let their body tissues outgas nitrogen slowly enough that bubbles are not formed. When the astronaut returns to the "shirt sleeve" environment of the spacecraft after an EVA, pressure is restored to whatever the operating pressure of that spacecraft may be, generally normal atmospheric pressure. Decompression illness in spaceflight consists of decompression sickness (DCS) and other injuries due to uncompensated changes in pressure, or barotrauma.

Decompression sickness

Decompression sickness is the injury to the tissues of the body resulting from the presence of nitrogen bubbles in the tissues and blood. This occurs due to a rapid reduction in ambient pressure causing the dissolved nitrogen to come out of solution as gas bubbles. In space the risk of DCS is significantly reduced by using a technique to wash out the nitrogen in the body's tissues. This is achieved by breathing 100% oxygen for a specified period of time before donning the spacesuit, and is continued after a nitrogen purge. DCS may result from inadequate or interrupted pre-oxygenation time, or other factors including the astronaut's level of hydration, physical conditioning, prior injuries and age. Other risks of DCS include inadequate nitrogen purge in the EMU, a strenuous or excessively prolonged EVA, or a loss of suit pressure. Non-EVA crewmembers may also be at risk for DCS if there is a loss of spacecraft cabin pressure.

Symptoms of DCS in space may include chest pain, shortness of breath, cough or pain with a deep breath, unusual fatigue, lightheadedness, dizziness, headache, unexplained musculoskeletal pain, tingling or numbness, extremities weakness, or visual abnormalities.

Primary treatment principles consist of in-suit repressurization to re-dissolve nitrogen bubbles, 100% oxygen to re-oxygenate tissues, and hydration to improve the circulation to injured tissues.

To date there have been no reported cases of DCS in the NASA space program.

Barotrauma

Barotrauma is the injury to the tissues of air filled spaces in the body as a result of differences in pressure between the body spaces and the ambient atmospheric pressure. Air filled spaces include the middle ears, paranasal sinuses, lungs and gastrointestinal tract. One would be predisposed by a pre-existing upper respiratory infection, nasal allergies, recurrent changing pressures, dehydration, or a poor equalizing technique.

Positive pressure in the air filled spaces results from reduced barometric pressure during the depressurization phase of an EVA. It can cause abdominal distension, ear or sinus pain, decreased hearing, and dental or jaw pain. Abdominal distension can be treated with extending the abdomen, gentle massage and encourage passing flatus. Ear and sinus pressure can be relieved with passive release of positive pressure. Pretreatment for susceptible individuals can include oral and nasal decongestants, or oral and nasal steroids.

Negative pressure in air fill spaces results from increased barometric pressure during repressurization after an EVA or following a planned restoration of a reduced cabin pressure. Common symptoms include ear or sinus pain, decreased hearing, and tooth or jaw pain.

Treatment may include active positive pressure equalization of ears and sinuses, oral and nasal decongestants, or oral and nasal steroids, and appropriate pain medication if needed.

Medicine in Space

The Advanced Diagnostic Ultrasound in Microgravity Study is funded by the National Space and Biomedical Research Institute and involves the use of ultrasound among Astronauts including former ISS Commanders Leroy Chiao and Gennady Padalka who are guided by remote experts to diagnose and potentially treat hundreds of medical conditions in space. This study has a widespread impact and has been extended to cover professional and Olympic sports injuries as well as medical students. It is anticipated that remote guided ultrasound will have application on Earth in emergency and rural care situations. Findings from this study were submitted for publication to the journal Radiology aboard the International Space Station; the first article submitted in space. [3] [4] [5]

References

Altitude Decompression Sickness Susceptibility, MacPherson, G; Aviation, Space, and Environmental Medicine, Volume 78, Number 6, June 2007 , pp. 630-631(2)

Decision Analysis in Aerospace Medicine: Costs and Benefits of a Hyperbaric Facility in Space, John-Baptiste, A; Cook, T; Straus, S; Naglie, G; et al. Aviation, Space, and Environmental Medicine, Volume 77, Number 4, April 2006 , pp. 434-443(10)

Incidence of Adverse Reactions from 23,000 Exposures to Simulated Terrestrial Altitudes up to 8900 m, DeGroot, D; Devine JA; Fulco CS; Aviation, Space, and Environmental Medicine, Volume 74, Number 9, September 2003 , pp. 994-997(4)

Decreased immune system functioning

Astronauts in space have weakened immune systems, which means that in addition to increased vulnerability to new exposures, viruses already present in the body—which would normally be suppressed—become active. In space, T-cells (a part of white blood cells that produces antibodies) do not reproduce properly. T-cells that do exist are less able to fight off infection. NASA research is measuring the change in the immune systems of its astronauts as well as performing experiments with T-cells in space.

Loss of balance

Leaving and returning to Earth's gravity causes "space sickness," dizziness, and loss of balance in astronauts. By studying how changes can affect balance in the human body—involving the senses, the brain, the inner ear, and blood pressure—NASA hopes to develop treatments that can be used on Earth and in space to correct balance disorders. Until then, NASA's astronauts must rely on a medication called Midodrine (an "anti-dizzy" pill that temporarily increases blood pressure) to help carry out the tasks they need to do to return home safely.

Loss of bone density

Unlike people with osteoporosis, astronauts eventually regain their bone density. After a 3-4 month trip into space, it takes about 2–3 years to regain lost bone density. New techniques are being developed to help astronauts recover faster. Research in the following areas holds the potential to aid the process of growing new bone:

- Diet and Exercise changes may reduce osteoporosis.
- Vibration Therapy may stimulate bone growth.
- Medication could trigger the body to produce more of the protein responsible for bone growth and formation.

Loss of muscle mass

In space, muscles in the legs, back, spine, and heart weaken and waste away because they no longer are needed to overcome gravity, just as people lose muscle when they age due to reduced physical activity. Astronauts rely on research in the following areas to build muscle and maintain body mass:

- Exercise may build muscle if at least two hours a day is spent doing resistance training routines.
- Hormone supplements (hGH) may be a way to tap into the body's natural growth signals.
- Medication may trigger the body into producing muscle growth proteins.

Orthostatic intolerance

In space, astronauts lose fluid volume—including up to 22% of their blood volume. Because it has less blood to pump, the heart will atrophy. A weakened heart results in low blood pressure and can produce a problem with "orthostatic tolerance," or the body's ability to send enough oxygen to the brain without fainting or becoming dizzy. "Under the effects of the earth's gravity, blood and other body fluids are pulled towards the lower body. When gravity is taken away or reduced during space exploration, the blood tends to collect in the upper body instead, resulting in facial edema and other unwelcome side effects. Upon return to earth, the blood begins to pool in the lower extremities again, resulting in orthostatic hypotension."[6]

Radiation effects

Soviet cosmonaut Valentin Lebedev, who spent 221 days in the orbit in 1982 (an absolute record for stay in Earth's orbit), lost his eyesight to progressive cataract. Lebedev stated: "I suffered from a lot of radiation in space. It was all concealed back then, during the Soviet years, but now I can say that I caused damage to my health because of that flight."[7]

Sleep disorders

Fifty percent of space shuttle astronauts take sleeping pills and still get two hours or less of sleep. NASA is researching two areas which may provide the keys to a better night's sleep, as improved sleep decreases fatigue and increases daytime productivity:

- Environmental cues and practices may be able to retrain the body's circadian rhythm.
- Medication could improve sleep by increasing the production of melatonin, a sleep hormone

Medical interventions

- Exercise to maintain muscle strength and function
- Sleep cap
- Medication, including hormone replacement therapy

How does one prepare for a career in space medicine?

Educational programs

- Naval Operations Medical Institute (NAMI) - Pensacola, Florida [8]
- U.S. Air Force School of Aerospace Medicine - Brooks City-Base, San Antonio, Texas [9]
- University of Texas Medical Branch at Galveston (UTMB) - Aerospace Medicine Residency Program [10]
- Wright State University School of Medicine - Dayton, Ohio [11]
- Vanderbilt Center for Space Physiology and Medicine [12] - (Dr. David Robertson)
- Space Studies Department at the University of North Dakota

Related degrees, areas of specialization, and certifications

- Aeromedical certification
- Aerospace Medicine
- Aerospace Studies
- Emergency Medicine
- Family Practice
- Internal Medicine
- Occupational and Preventive Medicine
- Ophthalmology
- Otolaryngology
- Public Health

Professional organizations

- Aerospace Human Factors Association [13]
- Aerospace Medical Association [14]
- Air Medical Physician Association [15]
- American Psychological Association [16]
- Association of Air Medical Services [17]
- Aviation & Aerospace Medicine [18]
- Federal Aviation Administration Office of Aerospace Medicine [19]
- Human Factors & Ergonomics Society [20]
- Human Performance in Extreme Environments [21]

See also

- Spaceflight osteopenia
- Artificial gravity
- Human adaptation to space
- Microgravity University
- Vomit Comet
- Health threat from cosmic rays

External links

- Description of space medicine [22]
- NASA History Series Publications [23] (many of which are online)

References

[1] BBC NEWS | Magazine | Project Paperclip: Dark side of the Moon (http://news.bbc.co.uk/1/hi/magazine/4443934.stm)
[2] Associated Press (2006-05-19). "Former Nazi removed from Space Hall of Fame" (http://msnbc.msn.com/id/12869190/). MSNBC. . Retrieved 2006-05-19.
[3] http://www.nasa.gov/mission_pages/station/science/experiments/ADUM.html
[4] A Pilot Study of Comprehensive Ultrasound Education at the Wayne State University School of Medicine: http://www.jultrasoundmed.org/cgi/content/abstract/27/5/745
[5] Evaluation of Shoulder Integrity in Space: First Report of Musculoskeletal US on the International Space Station: http://radiology.rsna.org/content/234/2/319.abstract
[6] When Space Makes You Dizzy (http://science.nasa.gov/headlines/y2002/25mar_dizzy.htm)
[7] Soviet cosmonauts burnt their eyes in space for USSR's glory (http://english.pravda.ru/science/earth/17-12-2008/106841-soviet_cosmonaut-0)
[8] http://www.nomi.med.navy.mil/
[9] http://www.brooks.af.mil/units/usafsam/index.asp
[10] http://www.utmb.edu/pmr/
[11] http://www.med.wright.edu/asm/res/asmhome.html
[12] http://www.mc.vanderbilt.edu/gcrc/space/
[13] http://www.asma.org/Organization/ashfa/
[14] http://www.asma.org
[15] http://www.ampa.org/
[16] http://www.apa.org
[17] http://www.aams.org/
[18] http://members.ozemail.com.au/~dxw/avmed.html
[19] http://www.cami.jccbi.gov/
[20] http://www.hfes.org
[21] http://www.hpee.org
[22] http://iss.sfo.jaxa.jp/med/index_e.html
[23] http://history.nasa.gov/series95.html

Space adaptation syndrome

Space adaptation syndrome (SAS) or **space sickness** is a condition experienced by around half of space travelers during adaptation to weightlessness.[2] It is related to motion sickness, as the vestibular system adapts to weightlessness.[3]

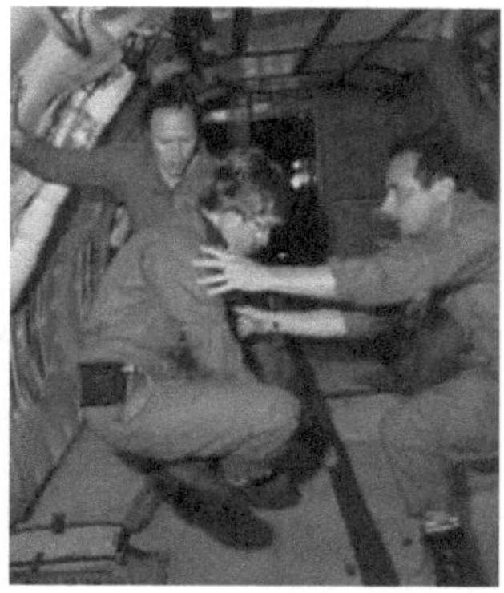

NASA astronauts acclimating themselves to space adaptation syndrome in a KC-135 airplane that flies parabolic arcs to create short periods of weightlessness.[1]

Space sickness is caused by changes in g-forces, which affect spatial orientation in humans.[3] According to *Science Daily*, "Gravity plays a major role in our spatial orientation. Changes in gravitational forces, such as the transition to weightlessness during a space voyage, influence our spatial orientation and require adaptation by many of the physiological processes in which our balance system plays a part. As long as this adaptation is incomplete, this can be coupled to motion sickness (nausea), visual illusions and disorientation."[3] One understanding of motion sickness is that nausea is pro-survival evolutionary adaptation, because the sensory stimulation of a maladapted high acceleration environment that the body is not accustomed to is recognized by the brain as being similar to the sensory conflict from eating a poisonous plant, in which case vomiting is a helpful reaction.

Space sickness was effectively unknown during the earliest spaceflights as these were undertaken in very cramped conditions; it seems to be aggravated by being able to freely move around and so is more common in larger spacecraft. Both Frank Borman, (*Apollo 8*) and Rusty Schweickart, (*Apollo 9*) experienced identifiable and fairly severe symptoms, in the latter case causing the mission plan to be modified. Neither astronaut was assigned to a mission after their SAS incident. Subsequent astronauts (e.g. the Skylab 4 crew) attempted to prevent Mission Control from learning about their own SAS experience, apparently out of concern for their future flight assignment potential.

As with motion sickness, symptoms can vary from mild nausea and disorientation, to vomiting and intense discomfort; headaches and nausea are often reported in varying degrees. About half of sufferers experience mild symptoms; only around 10% suffer severely. The most extreme reaction yet recorded was that felt by Senator Jake Garn in 1985. After his flight NASA astronauts began using the informal "Garn scale"[4] to measure reactions to space sickness. In most cases, symptoms last from 2–4 days.

Modern motion-sickness medications can counter space sickness but are rarely used because it is considered better to allow space travelers to adapt naturally over the first day or two than to suffer the drowsiness and other side effects of medication. However, transdermal dimenhydrinate anti-nausea patches are typically used whenever space suits are worn because vomiting into a space suit could be fatal. Space suits are generally worn during launch and landing by NASA crew members and always for extra-vehicular activities (EVAs). EVAs are consequently not usually scheduled for the first days of a mission to allow the crew to adapt, and transdermal dimenhydrinate patches are typically used as an additional backup measure.

See also

- Human adaptation to space
- Human spaceflight

References

[1] "Mixed Up in Space" (http://science.nasa.gov/headlines/y2001/ast07aug_1.htm). NASA. 2001-08-07. . Retrieved 2009-05-28.

[2] Quine, Tony (April 2007). "Addicted to space: An appreciation of Anousheh Ansari, Part II". *Spaceflight* (British Interplanetary Society (BIS)) **49** (4): 144. ISSN 0038-6340.

[3] "Why Do Astronauts Suffer From Space Sickness?" (http://www.sciencedaily.com/releases/2008/05/080521112119.htm). ScienceDaily. 2008-05-23. .

[4] http://www.jsc.nasa.gov/history/oral_histories/StevensonRE/RES_5-13-99.pdf, pg 35, Johnson Space Center Oral History Project, interview with Dr. Robert Stevenson:

> "Jake Garn was sick, was pretty sick. I don't know whether we should tell stories like that. But anyway, Jake Garn, he has made a mark in the Astronaut Corps because he represents the maximum level of space sickness that anyone can ever attain, and so the mark of being totally sick and totally incompetent is one Garn. Most guys will get maybe to a tenth Garn, if that high. And within the Astronaut Corps, he forever will be remembered by that."

Space tourism

Space tourism is tourism in which participants pay for flights into space. As of 2010, orbital space tourism opportunities are limited and expensive, with only the Russian Space Agency providing transport. The price for a flight brokered by Space Adventures to the International Space Station aboard a Soyuz spacecraft is US$ 20–35 million. The space tourists usually sign contracts with third parties to conduct certain research while in orbit. This helps to minimize their own expenses.

Space tourist Mark Shuttleworth

Infrastructure for a suborbital space tourism industry is being developed through the construction of spaceports in numerous locations, including California, Oklahoma, New Mexico, Florida, Virginia, Alaska, Wisconsin, Esrange in Sweden as well as the United Arab Emirates. Some use the term "personal spaceflight" as in the case of the Personal Spaceflight Federation.

A number of startup companies have sprung up in recent years, hoping to create a space tourism industry. For a list of such companies, and the spacecraft they are currently building, see list of space tourism companies.

Russia halted space tours since 2010, however it is planning to resume them in 2012, when the number of single-use three-man Soyuz launches rises to five a year.[1] [2] [3]

Early dreams

After early successes in space, much of the public saw intensive space exploration as inevitable. Those aspirations are remembered in science fiction such as Arthur C. Clarke's *A Fall of Moondust* and also *2001: A Space Odyssey*, Roald Dahl's *Charlie and the Great Glass Elevator*, Joanna Russ's 1968 novel *Picnic on Paradise*, and Larry Niven's Known Space stories. Lucian in the second century AD in his book *True History* examines the idea of a crew of men whose ship travels to the Moon during a storm. Jules Verne also took up the theme of lunar visits in his books, *From the Earth to the Moon* and *Around the Moon*. Robert A. Heinlein's short story *The Menace from Earth*, published in

1957, was one of the first to incorporate elements of a developed space tourism industry within its framework. During the 1960s and 1970s, it was common belief that space hotels would be launched by 2000. Many futurologists around the middle of the 20th century speculated that the average family of the early 21st century would be able to enjoy a holiday on the Moon. In the 1960s, Pan Am established a waiting list for future flights to the moon,[4] issuing free "First Moon Flights Club" membership cards to those who requested them.

The end of the Space Race, however, signified by the Moon landing, decreased the emphasis placed on space exploration by national governments and therefore led to decreased demands for public funding of manned space flights.[5]

Precursors

The Soviet space program was aggressive in broadening the pool of cosmonauts. The Soviet Intercosmos program also included cosmonauts selected from Warsaw Pact members (from Czechoslovakia, Poland, East Germany, Bulgaria, Hungary, Romania) and later from allies of the USSR (Cuba, Mongolia, Vietnam) and non-aligned countries (India, Syria, Afghanistan). Most of these cosmonauts received full training for their missions and were treated as equals, but especially after the Mir program began, were generally given shorter flights than Soviet cosmonauts. The European Space Agency took advantage of the program as well.

The U.S. space shuttle program included payload specialist positions which were usually filled by representatives of companies or institutions managing a specific payload on that mission. These payload specialists did not receive the same training as professional NASA astronauts and were not employed by NASA. In 1983, Ulf Merbold from ESA and Byron Lichtenberg from MIT (engineer and Air Force fighter pilot) were the first Payload Specialists to fly on the shuttle, becoming the first non-NASA astronauts. In 1984, Charlie Walker became the first non-government astronaut to fly, with his flight paid for by his employer, McDonnell Douglas. NASA was also eager to prove its capability to Congressional sponsors. Senator Jake Garn was flown on the shuttle in 1985, followed by Representative Bill Nelson (now Senator) in 1986. As the shuttle program expanded, the Teacher in Space program was developed as a way to expand publicity and educational opportunities for NASA. Christa McAuliffe would have been the first Teacher in Space, but was killed in the Challenger disaster and the program was canceled. During the same period a Journalist in Space program was frequently discussed, with individuals such as Walter Cronkite considered a front-runners, but no formal program was ever developed. McAuliffe's backup in the Teacher in Space Program, Barbara Morgan, eventually got hired in 1998 as a professional astronaut and flew on STS-118 as a mission specialist where she spoke to many students as an educator during the trip.

A second journalist-in-space program, in which NASA green-lighted Miles O'Brien to fly on the space shuttle, was scheduled to be announced in 2003. That program was canceled in the wake of the Columbia accident on STS-107 and subsequent emphasis on finishing the International Space Station before retiring the space shuttle.

With the realities of the post-Perestroika economy in Russia, its space industry was especially starved for cash. The Tokyo Broadcasting System (TBS) offered to pay for one of its reporters to fly on a mission. For $28 million, Toyohiro Akiyama was flown in 1990 to Mir with the eighth crew and returned a week later with the seventh crew. Akiyama gave a daily TV broadcast from orbit and also performed scientific experiments for Russian and Japanese companies. However, since the cost of the flight was paid by his employer, Akiyama could be considered a business traveler rather than a tourist.

In 1991, British chemist Helen Sharman was selected from a pool of 13,000 applicants to be the first Briton in space.[6] The program was known as Project Juno and was a cooperative arrangement between the Soviet Union and a group of British companies. The Project Juno consortium failed to raise the funds required, and the program was almost cancelled. Reportedly Mikhail Gorbachev ordered it to proceed under Soviet expense in the interests of international relations, but in the absence of Western underwriting, less expensive experiments were substituted for those in the original plans. Sharman flew aboard Soyuz TM-12 to Mir and returned aboard Soyuz TM-11.

Orbital space tourism

At the end of the 1990s, MirCorp, a private venture by then in charge of the space station, began seeking potential space tourists to visit Mir in order to offset some of its maintenance costs. Dennis Tito, an American businessman and former JPL scientist, became their first candidate. When the decision to de-orbit Mir was made, Tito managed to switch his trip to the International Space Station through a deal between MirCorp and U.S.-based Space Adventures, Ltd., despite strong opposition from senior figures at NASA. Space Adventures remains the only company to have sent paying passengers to space.[7] [8]

In conjunction with the Federal Space Agency of the Russian Federation and Rocket and Space Corporation (Energia), Space Adventures facilitated the flights for all of the world's first private space explorers. The first three participants paid in excess of $20 million (USD) each for their 10-day visit to the ISS.

On April 28, 2001, Dennis Tito became the first "fee-paying" space tourist when he visited the International Space Station (ISS) for seven days. He was followed in 2002 by South African computer millionaire Mark Shuttleworth. The third was Gregory Olsen in 2005, who was trained as a scientist and whose company produced specialist high-sensitivity cameras. Olsen planned to use his time on the ISS to conduct a number of experiments, in part to test his company's products. Olsen had planned an earlier flight, but had to cancel for health reasons.

After the Columbia disaster, space tourism on the Russian Soyuz program was temporarily put on hold, because Soyuz vehicles became the only available transport to the ISS. However, in 2006, space tourism was resumed (Why?). On September 18, 2006, Anousheh Ansari, an Iranian American (Soyuz TMA-9), became the fourth space tourist (she prefers "private space explorer".[9]). On April 7, 2007, Charles Simonyi, an American businessman of Hungarian descent, joined their ranks (Soyuz TMA-10).

In 2003, NASA and the Russian Space Agency agreed to use the term 'Spaceflight Participant' to distinguish those space travelers from astronauts on missions coordinated by those two agencies. Tito, Shuttleworth, Olsen, Ansari, and Simonyi were designated as such during their respective space flights.[10] NASA also lists Christa McAuliffe as a "Space Flight Participant" (although she did not pay a fee), apparently due to her non-technical duties aboard the STS-51-L flight.

As reported by Reuters on 3 March 2010, Russia announced that the country will double the number of launches of three-man Soyuz ships to four that year, because "permanent crews of professional astronauts aboard the expanded [ISS] station are set to rise to six"; regarding space tourism, the head of the Russian Cosmonauts' Training Center said "for some time there will be a break in these journeys".[1] Then, on 18 March 2010 Interfax news agency reported that Russia plans to launch five Soyuz spacecrafts a year since 2012: four per the ISS program, and an additional spacecraft which may be used for space tourism.[2]

List of flown space tourists

Seven of the space tourists flew to and from the International Space Station on Soyuz spacecraft through the space tourism company, Space Adventures:[11]

Space tourist	Nationality	Year	Duration of flight	Flight
Dennis Tito	American	2001	9 days (Apr 28 – May 6)	Launch: Soyuz TM-32 Return: Soyuz TM-31
Mark Shuttleworth	South African	2002	11 days (Apr 25 – May 5)	Launch: Soyuz TM-34 Return: Soyuz TM-33
Gregory Olsen	American	2005	11 days (Oct 1 – Oct 11)	Launch: Soyuz TMA-7 Return: Soyuz TMA-6
Anousheh Ansari	Iranian / American	2006	12 days (Sept 18 – Sept 29)	Launch: Soyuz TMA-9 Return: Soyuz TMA-8
Charles Simonyi[12]	Hungarian / American	2007	15 days (Apr 7 – Apr 21)	Launch: Soyuz TMA-10 Return: Soyuz TMA-9
		2009	14 days (Mar 26 – Apr 8)	Launch: Soyuz TMA-14 Return: Soyuz TMA-13
Richard Garriott[13]	American / British	2008	12 days (Oct 12 – Oct 23)	Launch: Soyuz TMA-13 Return: Soyuz TMA-12
Guy Laliberté	Canadian	2009	12 days (Sept 30 – Oct 11)	Launch: Soyuz TMA-16 Return: Soyuz TMA-14

Suborbital flights

More affordable suborbital space tourism is viewed as a money-making proposition by several other companies, including Space Adventures, Virgin Galactic, Starchaser, Blue Origin, Armadillo Aerospace, XCOR Aerospace, Rocketplane Limited, the European "Project Enterprise",[14] and others. Most are proposing vehicles that make suborbital flights peaking at an altitude of 100-160 kilometres.[15] Passengers would experience three to six minutes of weightlessness, a view of a twinkle-free starfield, and a vista of the curved Earth below. Projected costs are expected to be about $200,000 per passenger.[16] [17] [18] [19] [20]

As of November 2007 Virgin Galactic had pre-sold nearly 200 seats for their suborbital space tourism flights, according to the company's president.[21]

The X Prize

The X-Prize being awarded to the Scaled Composites team

On October 4, 2004, the SpaceShipOne, designed by Burt Rutan of Scaled Composites, won the $10,000,000 X Prize, which was designed to be won by the first private company who could reach and surpass an altitude of 62 miles (100 km) twice within two weeks. The altitude is beyond the Kármán Line, the arbitrarily defined boundary of space.[22] The first flight was flown by Michael Melvill on June 21, 2004 to a height of 62 miles, making him the first commercial astronaut.[23] The prize-winning flight was flown by Brian Binnie, which reached a height of 69.6 miles, breaking the X-15 record.[22]

Virgin Galactic

Spaceship One, the first privately funded and constructed spacecraft to fly above the 100 km Karman Line.

Virgin Galactic, one of the leading potential space tourism groups, is planning to begin passenger service aboard the VSS Enterprise, a Scaled Composites SpaceShipTwo type spacecraft. The initial seat price will be $200,000, but that price is expected to eventually fall to $20,000. To date, over 80,000 people have made down payments on bookings. Headed by Sir Richard Branson's Virgin Group, Virgin Galactic hopes to be the first private space tourism company to regularly send civilians into space. A citizen astronaut will only require three days of training before spaceflight. SpaceShipTwo is a scaled up version of SpaceShipOne, the spacecraft which claimed the Ansari X Prize. Both spacecraft were designed by Burt Rutan's Scaled Composites. Launches will initially occur at the Mojave Spaceport in California, and will then be moved to Spaceport America in Upham, New Mexico. Tourists will also be flown from Kiruna, Sweden[24] The spacecraft will travel 360,000 feet (109.73 km/68.18 miles) high. This goes beyond the internationally defined boundary between Earth and space of 100 km. Spaceflights will last 2.5 hours, carry 6 passengers, and reach a speed of Mach 3. SpaceShipTwo will not require a space shuttle-like heat shield for atmospheric reentry as it will not experience the extreme aerodynamic heating experienced during reentry at orbital velocities (approximately Mach 22.5 at a typical shuttle altitude of 300 km, or 185 miles).[25] The glider will employ a "feathering" technique to manage drag during the unpowered descent and landing. SpaceShipTwo will use a single hybrid rocket motor to launch from mid-air after detaching from a mother ship at 50,000 feet, instead of NASA's space shuttle's ground-based launch.

Project Enterprise

Project Enterprise was launched by the German TALIS Institute in 2004 and is the first project of its kind in Europe.[26] The goal is to develop a rocket-propelled spaceplane by 2011 that will carry one pilot and up to five passengers into suborbital space. The plane will launch from the ground using rockets, and will return in an unpowered flight like Virgin Galactic's SpaceShipTwo. The prototypes and finished spaceplane will be launched from an airport near Cochstedt (Germany; Saxony-Anhalt).

Since 2004, the TALIS Institute has gained many industrial partners, including XtremeAir,[27] who will manufacture the airframe, and Swiss Propulsion Laboratory SPL,[28] who will deliver the propulsion components. XtremeAir is known for their acrobatic airplanes, and SPL has designed and tested liquid propellant rocket engines since 1998.

Current work is focusing on the first prototype, the "Black Sky": An existing acrobatic airplane that would be fitted with a single rocket engine and a new wing. The rocket engine is expected to deliver a thrust of 10 kN. The test program for this engine started in 2007 at SPL and is expected to fly by 2010.[29] [30]

Legality

United States

In December 2005, the U.S. Government released a set of proposed rules for space tourism.[31] These included screening procedures and training for emergency situations, but not health requirements.

Under current US law, any company proposing to launch paying passengers from American soil on a suborbital rocket must receive a license from the Federal Aviation Administration's Office of Commercial Space Transportation (FAA/AST). The licensing process focuses on public safety and safety of property, and the details can be found in the Code of Federal Regulations, Title 14, Chapter III.[32] This is in accordance with the Commercial Space Launch

Amendments Act passed by Congress in 2004.[33]

Orbital flights, space stations and space hotels

- EADS Astrium, a subsidiary of European aerospace giant EADS, announced its space tourism project on June 13, 2007.[34]
- SpaceX is a private space company which is developing their own rocket family called *Falcon* and a capsule named Dragon, capable of sending up to 7 people to any space station, either ISS or a possible station by Bigelow Aerospace. Falcon 1 has already undertaken testflights and successfully completed its first commercial flight on July 14, 2009, deploying the Malaysian RazakSAT into orbit. Falcon 9 (which will be the rocket for the Dragon capsule) was first launched June 4, 2010 at Space Launch Complex 40 in Cape Canaveral. An initial prototype of the Dragon capsule is expected to be used on this test flight; SpaceX anticipates that Dragon could be qualified for human spaceflight within 3 years of the receipt of NASA CCDV [35] funding.[36]
- Space Adventures Ltd. have also announced that they are working on circumlunar missions to the moon, with the price per passenger being $100,000,000.[37] They are currently developing spaceports at the United Arab Emirates (Ras al-Khaimah) and in Singapore.
- Orbital space tourist flights are also being planned by Excalibur Almaz, using modernized *TKS* space capsules.[38]

Several plans have been proposed for using a space station as a hotel.[39] American motel tycoon Robert Bigelow has acquired the designs for inflatable space habitats from the Transhab program abandoned by NASA. His company, Bigelow Aerospace, has already launched two first inflatable habitat modules. The first, named Genesis I, was launched 12 July 2006. The second test module, Genesis II, was launched 28 June 2007. Both Genesis habitats remain in orbit as of mid-2009. As of 2006, Bigelow planned to officially launch the first commercial space station by 2012 (tagged *Nautilus*) which will have 330 cubic meters (almost as big as the ISS's 425 cubic meters of usable volume).[40] [41]

Bigelow Aerospace is currently offering the America's Space Prize, a $50 million prize to the first US company to create a reusable spacecraft capable of carrying passengers to a Nautilus space station.

Other companies have also expressed interest in constructing "space hotels". For example, Excalibur Almaz plans to modernize and launch its Soviet-era Almaz space stations, which will feature the largest windows ever on spacecraft. Virgin's Richard Branson has expressed his hope for the construction of a space hotel within his lifetime. He expects that beginning a space tourism program will cost $100 million.[16] Hilton International announced the Space Islands Project, a plan to connect together used space shuttle fuel tanks, each the diameter of a Boeing 747 aircraft.[42] A separate organization, Space Island Group[43] announced their distinct Space Island Project (note the singular "Island"), and plans on having 20,000 people on their "space island" by 2020, with the number of people doubling for each decade.[44] British Airways has expressed interest in the venture. If and when Space Hotels develop, it would initially cost a passenger $60,000, with prices lowering over time.[45]

Fashion designer Eri Matsui has designed clothing, including a wedding gown, intended to look best in weightless environments.

Advocacy, education, and industry organizations

Several organizations have been formed to promote the space tourism industry, including the Space Tourism Society, and others.[46] [47] [48] More information about the future of Space Tourism can be found at Space Tourism Lecture [49], which is a free online Space Tourism Lecture handout collection. Since 2003 Dr. Robert A. Goehlich and Pierluigi Polignano teaches the world's first and only Space Tourism class at Keio University, Yokohama, Japan. UniGalactic Space Travel Magazine is a bi-monthly educational publication covering space tourism and space exploration developments in companies like SpaceX, Orbital Sciences, Virgin Galactic and organizations like NASA. The content of UniGalactic Space Travel Magazine can be found on UniGalactic web site. [50]

Opinions of commercial space tourism

A web-based survey suggested that over 70% of those surveyed wanted less than or equal to 2 weeks in space; in addition, 88% wanted to spacewalk (only 74% of these would do it for a 50% premium), and 21% wanted a hotel or space station.[51]

The concept has met with some criticism from bureaucrats, notably Günter Verheugen, vice-president of the European Commission, who said of the EADS Astrium Space Tourism Project "It's only for the super rich, which is against my social convictions".[52]

Objections to "space tourist" terminology

Dennis Tito, Mark Shuttleworth, Gregory Olsen, Anousheh Ansari and Richard Garriott have all expressed a preference to be called something other than "space tourist". In each case, they explained their preferences by pointing out that they carried out scientific experiments as part of their journey; Garriott additionally emphasized their training is identical to requirements of non-Russian Soyuz crew members, and that teachers and other non-professional astronauts chosen to fly with NASA are called astronauts.[53] Garriott prefers "cosmonaut" or "astronaut", but will accept "private" in front of either. Tito has asked to be known as an "independent researcher". Shuttleworth proposed "pioneer of commercial space travel".[54] Olsen preferred "private researcher."[55] Ansari prefers the term "private space explorer".[9] Alone among those who have paid to go to orbit so far, Charles Simonyi seems to have no concerns about calling it "space tourism", even in reference to his own experience. Asked in an interview "Do you foresee a day when space tourism is not just the province of billionaires - when it will be as affordable as plane travel?", he did not object to the implicit categorization of his own trip, but rather answered "Yes, the only question is when ...."[56]

NASA and the Russian Federal Space Agency agreed to use the term "spaceflight participant" to distinguish those space travelers from astronauts on missions coordinated by those two agencies.

Although many space enthusiasts subscribe to the notion of space tourism as a potential burgeoning industry that could further the development and settlement of space, some of these same enthusiasts object to the use of the term "space tourist". Rick Tumlinson of the Space Frontier Foundation, for example, has said

> "I hate the word tourist, and I always will .... 'Tourist' is somebody in a flowered shirt with three cameras around his neck."[57]

Others with perhaps less enthusiasm for space development seem to agree. Alex Tabarrok has categorized it as a kind of "adventure travel". The mere fact of people paying for a travel experience does not, in his view, make that activity "tourism".

> "At best and for the foreseeable future space travel will remain akin to climbing Everest, dangerous and uncommon. Yes, we might see 100 flights a year but that's not space tourism - tourism is fat guys with cameras."[58]

Brian Binnie, and Mike Melvill, the pilots of Scaled Composites Space Ship One were awarded the title of Commercial Astronaut by the United States Federal Aviation Administration. This is a new classification that distinguishes the holder as an astronaut, but is not associated with the United States NASA space program.

See also

- Commercialization of space
- Commercial astronaut
- Launch loop
- Quasi Universal Intergalactic Denomination
- Space colonization
- Space Tourism Society
- Spaceflight participant
- Sex in space
- Space architecture
- Timeline of Solar System exploration

Businesses

- List of private spaceflight companies
- Armadillo Aerospace
- Bigelow Aerospace
- Blue Origin
- MirCorp
- Space Adventures
- Space Transport Corporation (STC)
- Virgin Galactic

External links

- Space travel guide from Wikitravel
- Space Tourists (A documentary film by Christian Frei) [59]
- Selling Peace (A book about early space tourism and privatization) [60]
- Galactic Suite Space Hotel [61]
- Galactic Suite Design Aerospace Experiences [61]
- Bloon Near-Space travel [62]

References

[1] "Russia halts space tours as U.S. retires Shuttle" (http://www.reuters.com/article/idUSTRE6223VF20100303). 2010-03-03. .
[2] Russia will restore space tourism since 2012 (http://interfax.ru/society/news.asp?id=128393) (in Russian)
[3] Russia to resume space tourism in 2012 (http://adm.rt.tv/Top_News/2010-03-18/russia-space-tourist-resume.html), RT, March 18, 2010
[4] Beveridge, Dirk (1991-07-05). "Pan Am lunar list tethered to earth" (http://news.google.com/newspapers?id=-7wMAAAAIBAJ&sjid=dl8DAAAAIBAJ&pg=5479,3673315&dq=pan-am+space-odyssey). *St. Petersburg Times*. . Retrieved 2009-06-01.
[5] *Space: the forgotten frontier?* (http://www.oecdobserver.org/news/fullstory.php/aid/963/Space:_the_forgotten_frontier_.html). 2003-03-01. .
[6] "1991: Sharman becomes first Briton in space (http://news.bbc.co.uk/onthisday/hi/dates/stories/may/18/newsid_2380000/2380649.stm)"]. *BBC News* (BBC News Online). 1991-05-18. . Retrieved 2010-01-04.
[7] "Int'l space station ticket price climbs" (http://abcnews.go.com/Technology/wireStory?id=3391085). AP. .
[8] Boyle, Alan (June 13, 2006). "Regulators OK Oklahoma spaceport - Suborbital test flights could begin in 2007, setting stage for tourists" (http://msnbc.msn.com/id/13304491). MSNBC. . Retrieved 2006-06-26.
[9] "www.anoushehansari.com" (http://www.anoushehansari.com/). . Retrieved 2006-09-22.
[10] "Payload Specialist Astronauts" (http://www.jsc.nasa.gov/Bios/PS/index.html). . Retrieved 2007-05-20.
[11] howstuffworks.com. "How Space Tourism Works" (http://science.howstuffworks.com/space-tourism.htm). .
[12] Charles in Space (http://www.charlesinspace.com/) Charles Simonyi's blog and video blog about his trip to the ISS
[13] NASA (2008). "Expedition 18" (http://www.nasa.gov/mission_pages/station/expeditions/expedition18/index.html). NASA. . Retrieved September 26, 2008.

[14] "Project Enterprise" (http://www.european-spacetourism.eu). . Retrieved 2008-03-26.
[15] *Scotland 2040: Spaceships head for Moon with lunar golfers and crater ramblers aboard* (http://news.scotsman.com/topics.cfm?tid=6&id=1510152006). 2006-10-12. .
[16] *British tycoon wants to fly you to space: Virgin Galactic plans to sell $200,000 rides* (http://www.sfgate.com/cgi-bin/article.cgi?file=/c/a/2004/09/28/MNG99906EQ1.DTL). San Francisco Chronicle. 2004-09-28. .
[17] "X PRIZE Foundation" (http://www.xprize.org/). .
[18] "Space Adventures" (http://www.spaceadventures.com/). .
[19] "Zero Gravity Corporation" (http://www.nogravity.com/). .
[20] The Last Frontier Of Tourism (http://www.monochrom.at/space-tourism/) (article by Stefan Tiron, published by monochrom)
[21] "Spaceflights now for sale; scary part is price" (http://seattlepi.nwsource.com/business/339535_spacetourism14.html). Seattle Post-Intelligencer. November 14, 2007. .
[22] SpaceShipOne Captures X-Prize (http://www.scaled.com/projects/tierone/041004_spaceshipone_x-prize_flight_2.html) October 4, 2004
[23] SpaceShipOne pilot bio: Michael W. Melvill (http://www.scaled.com/projects/tierone/melvill.htm) May 2005
[24] (http://www.ssc.se/?id=9500) — see also Spaceport Sweden
[25] *Flight to Orbit* (http://exploration.grc.nasa.gov/education/rocket/rktrflght.html). .
[26] european-spacetourism.eu. "Project Enterprise" (http://www.european-spacetourism.eu) (in German). european-spacetourism.eu. . Retrieved 2008-03-26.
[27] "Project Enterprise - XtremeAir" (http://www.xtremeair.de). . Retrieved 2008-03-26.
[28] "Project Enterprise - Swiss Propulsion Laboratory SPL" (http://www.spl.ch). . Retrieved 2008-03-26.
[29] Die Top Ten der privaten Raumfahrt (2) (http://www.der-orion.com/index.php?option=com_content&task=view&id=187&Itemid=1) (German)
[30] Project Enterprise new debut (http://www.hobbyspace.com/nucleus/index.php?itemid=5425)
[31] US draws up space tourism rules (http://news.bbc.co.uk/2/low/science/nature/4589072.stm)
[32] *Electronic Code of Federal Regulations* (http://ecfr.gpoaccess.gov/cgi/t/text/text-idx?sid=f32ec318140f194f1e3f1981d8192833&c=ecfr&tpl=/ecfrbrowse/Title14/14cfrv4_02.tpl#300). .
[33] *Congress Passes Space Tourism Bill* (http://www.msnbc.msn.com/id/6682611/). MSNBC. 2004-12-23. .
[34] Europe joins space tourism race (http://www.timesonline.co.uk/tol/news/world/europe/article1909818.ece), *Times online*, June 10, 2007
[35] http://procurement.jsc.nasa.gov/ccdev/
[36] Official SpaceX Updates (http://www.spacex.com/updates.php), Update from 2009-09-23.
[37] *$100 Million Moon Trip: Space Tourism's Hot Ticket?* (http://news.nationalgeographic.com/news/2005/08/0810_050810_moontrip.html). 2005-08-10. .
[38] http://www.spaceflightnow.com/news/n0908/18almaz/
[39] "Space hotels" (http://www.spacefuture.com/tourism/hotels.shtml). .
[40] *A Room with a View of Mars, Please* (http://www.tfot.info/content/view/77/). 2006-10-06. .
[41] "An interview with Michael Gold about Bigelow Aerospace" (http://www.tfot.info/content/view/77/). .
[42] "Hilton to back space hotel" (http://news.bbc.co.uk/1/hi/sci/tech/293366.stm). *BBC News*. 1999-03-09. . Retrieved 2007-05-20.
[43] "The Space Island Group's Mission" (http://www.spaceislandgroup.com/home.html). 2006. . Retrieved 2007-05-20.
[44] http://www.spaceislandgroup.com/sig-vision.html.
[45] *Space Future - Prospects of Space Tourism* (http://www.spacefuture.com/archive/prospects_of_space_tourism.shtml)*]. 1996=05-15.* .
[46] Space Tourism Society (http://spacetourismsociety.org/)
[47] "Space Future Journal" (http://www.spacefuture.com/). .
[48] *Space Tourism: Personal Spaceflight for you ...* (http://www.hobbyspace.com/Tourism). .
[49] http://www.robert-goehlich.de
[50] http://www.unigalactic.com
[51] http://www.space.com/news/061003_tourism_survey.html.
[52] "EU official balks at space tourism" (http://www.thepeninsulaqatar.com/Display_news.asp?section=World_News&subsection=Rest+of+the+World&month=June2007&file=World_News2007061523649.xml). 2007-06-15. .
[53] "Do Not Call Me A Space Tourist" (http://www.richardinspace.com/index.cfm?fuseaction=blog.welcome&theyear=2008&themonth=1&view=18). . Retrieved 2008-12-02.
[54] ""Space tourist" insists on pioneering role" (http://www.newscientist.com/article/dn2190-space-tourist-insists-on-pioneering-role.html). New Scientist. . Retrieved 2008-05-03.
[55] "Space millionaire to mix science with pleasure publisher= MSNBC" (http://www.msnbc.msn.com/id/4623674/). . Retrieved 2008-06-03.
[56] Arango, Tim (2007-05-25). "From the Iron Curtain to the Final Frontier" (http://money.cnn.com/magazines/fortune/fortune_archive/2007/05/28/100033715/?postversion=2007052514). *Fortune magazine*. . Retrieved 2008-06-05.
[57] Foust, Jeff. "Is it time to dump the t-word?" (http://www.thespacereview.com/article/280/1). . Retrieved 2008-06-03.

[58] Tabarrok, Alex. "Romance and Realism in Space Tourism" (http://www.marginalrevolution.com/marginalrevolution/2004/11/space_tourism_i.html). . Retrieved 2008-06-03.
[59] http://www.space-tourists-film.com
[60] http://www.sellingpeace.com/
[61] http://www.galacticsuite.com
[62] http://www.inbloon.com

Spaceflight participant

Spaceflight participant (Russian: Участник космического полёта, *uchastnik kosmicheskogo polyota*) is the term used by National Aeronautics and Space Administration (NASA) and the Russian Federal Space Agency (RKA) for people who travel aboard space missions coordinated by those agencies who are not part of the crew. The term serves to distinguish tourists and other special travelers from the career astronauts.

While the term gained new prominence with the rise of space tourism, it has also been used for participants in NASA's Teacher in Space program and for people who flew through inter-government agreements such as the Angkasawan program and the Korean Astronaut Program.

Other terms used for space travelers who are not career astronauts include NASA's Payload Specialist and the RKA's Researcher-Cosmonaut.

Background

The Soviet Intercosmos program included participants selected from Warsaw Pact members and later from allies of the USSR and non-aligned countries. Most of these people received full training for their missions and were treated as equals, but especially after the Mir program began, were generally given shorter flights than Soviet cosmonauts. The European Space Agency took advantage of the program as well.

The U.S. Space Shuttle program included payload specialist positions which were usually filled by representatives of companies or institutions managing a specific payload on that mission. These payload specialists did not receive the same level of training as career NASA astronauts and were not employed by NASA, so they were essentially private astronauts.

In the early days of the Shuttle program, NASA was also eager to prove its capability to Congressional sponsors, and Senator Jake Garn and (then-Representative, now Senator) Bill Nelson were both given opportunities to fly on board a Shuttle mission.

As the Shuttle program expanded, the Teacher in Space program was developed as a way to expand publicity and educational opportunities for NASA. Christa McAuliffe would have been the first Teacher in Space, but she was killed in the Challenger disaster and the program was canceled. During the same period a Journalist in Space program was frequently discussed, with individuals such as Walter Cronkite and Miles O'Brien considered front-runners, but no formal program was ever developed.[1] [2]

With the realities of the post-Perestroika economy in Russia, its space industry was especially starved for cash. The Tokyo Broadcasting System (TBS) offered to pay for one of its reporters to fly on a mission. For $28 million, Toyohiro Akiyama, was flown in 1990 to Mir with the eighth crew and returned a week later with the seventh crew. Akiyama gave a daily TV-broadcast from orbit and also performed scientific experiments for Russian and Japanese companies.

Since then, the Russian Federal Space Agency has also sold seats to a consortium of British companies for Project Juno, to six self-funded space tourists, to the Malaysian government as part of a contract to sell military planes, and to the South Korean government as part of the Korean Astronaut Program.

List of spaceflight participants

Name	Nationality	Program/Sponsor	Flight	Date	Comments
Christa McAuliffe[3]	United States	Teacher in Space Project	STS-51-L	28 January 1986	Killed alongside six fellow crew members in the Space Shuttle Challenger disaster. Back-up was Barbara Morgan, who was selected in 1998 to train as a Mission Specialist. Morgan finally flew to space aboard STS-118 in 2007, but as a "teacher-turned-astronaut", not a spaceflight participant.
Dennis Tito	United States	Self-funded space tourist	Soyuz TM-32 / Soyuz TM-31	April 28 - May 6, 2001	First space tourist
Mark Shuttleworth	South Africa	Self-funded space tourist	Soyuz TM-34 / Soyuz TM-33	April 25 - May 5, 2002	Shuttleworth was the first person with South African citizenship to fly in space.
Lance Bass	United States	Corporate-funded space tourist			Completed training but seat on Soyuz TMA-1 in 2002 was cancelled after funding fell through.
Gregory Olsen	United States	Self-funded space tourist	Soyuz TMA-7 / Soyuz TMA-6	October 1–11, 2005	
Daisuke Enomoto	Japan	Self-funded space tourist			Expected to fly on Soyuz TMA-9 in September 2006, but was grounded for medical reasons and seat was given to Ansari.
Anousheh Ansari	Iran / United States	Self-funded space tourist	Soyuz TMA-9 / Soyuz TMA-8	September 18–29, 2006	Trained as back-up to Enomoto. Was the first person with Iranian citizenship to fly in space.
Charles Simonyi	Hungary / United States	Self-funded space tourist	Soyuz TMA-10 / Soyuz TMA-9	April 7–21, 2007	
Sheikh Muszaphar Shukor	Malaysia	Angkasawan program	Soyuz TMA-11 / Soyuz TMA-10	10–21 October 2007	Back-up was Faiz Khaleed.
Yi So-yeon	South Korea	Korean Astronaut Program	Soyuz TMA-12 / Soyuz TMA-11	8–19 April 2008	Back-up was Ko San.
Richard Garriott	United States	Self-funded space tourist	Soyuz TMA-13 / Soyuz TMA-12	12 October 2008 - 23 October 2008	Back-up was Nik Halik.[4] [5]
Charles Simonyi	Hungary / United States	Self-funded space tourist	Soyuz TMA-14 / Soyuz TMA-13	26 March 2009 - 8 April 2009	Backup was Esther Dyson.[6] Simonyi was the first repeat space tourist.

Guy Laliberté	Canada	Self-funded space tourist	Soyuz TMA-16 / Soyuz TMA-14	30 September 2009 - 11 October 2009	First Canadian space tourist.[7] Backup was Barbara Barrett[8]
Upcoming					
Vladimir Gruzdev	Russia	Political party-sponsored trip			Expected to fly in 2009. The United Russia political party may pay the estimated $25 million for the flight from the party funds.[9]

All eight space tourism trips went to and from the International Space Station on Soyuz spacecraft and were arranged through the space tourism company, Space Adventures.[10]

Other missions

While not labeled as "spaceflight participants", the following people participated in spaceflight missions under the auspices of special programs outside of the professional astronaut corps.

Name	Nationality	Program/Sponsor	Flight	Date	Comments
Jake Garn	United States	US Government	STS-51-D	12–19 April 1985	To demonstrate the capabilities of the Space Shuttle, NASA offered a seat to Garn, a member of the Senate Appropriations Committee.[11]
Bill Nelson	United States	US Government	STS-61-C	12–18 January 1986	NASA also provided a seat to Nelson, a member of the U.S. House of Representatives.[12] He was originally scheduled to be aboard STS-51-L.
Edward C. Aldridge, Jr.	United States	US Government	STS-62-A		NASA assigned a seat to Aldridge, the Secretary of the Air Force, on mission STS-62-A, the first Shuttle mission scheduled to launch from Vandenberg Air Force Base.[13] After the Challenger disaster, the mission was cancelled and Aldridge never flew.
Toyohiro Akiyama	Japan	Tokyo Broadcasting System	Soyuz TM-11 / Soyuz TM-10	2–10 December 1990	As an employee of TBS, Akiyama could be considered the first space business traveler.
Helen Sharman	United Kingdom	Project Juno	Soyuz TM-12 / Soyuz TM-11	18–26 May 1991	Through Project Juno, a consortium of British companies partially funded a seat on a Soyuz flight to Mir (the Soviet Union covered the rest of the cost) in order to put the first Briton into space.[14]

See also

- Space tourism
- Intercosmos program
- Commercial astronaut

External links

- Charles in Space [15] Charles Simonyi's blog and video blog about his trip to the ISS.

References

[1] Roger Rosenblatt (2001-06-24). "A Realm Where Age Doesn't Count" (http://www.time.com/time/magazine/article/0,9171,139766,00.html?iid=chix-sphere). Time Magazine / CNN. . Retrieved September 12, 2007.

[2] Space Today Online (2005). "May fly sometime:" (http://www.spacetoday.org/Astronauts/BarbaraMorganTeacherAstronaut.html). Space Today Online. . Retrieved September 12, 2007.

[3] Payload Specialist Biographies (http://www.jsc.nasa.gov/Bios/PS/index.html#SFP)

[4] Tariq Malik (2008). "'Thrillionaire' Signs on as Backup Space Tourist" (http://news.yahoo.com/s/space/20080128/sc_space/thrillionairesignsonasbackupspacetourist). SPACE.com. . Retrieved January 28, 2008.

[5] ap.google.com, US game designer blasts into space with DNA cargo (http://ap.google.com/article/ALeqM5gVmXiJGB7S--M-E06nd_2DUW5sqQD93P1L181)

[6] U.S. Billionaire to Make Second Private Spaceflight (http://news.yahoo.com/s/space/20080930/sc_space/usbillionairetomakesecondprivatespaceflight)

[7] (http://news.yahoo.com/s/space/20090604/sc_space/acrobattobenextspacetourist)

[8] Жена главы компании Intel готовится стать космонавтом (http://www.rosbalt.ru/2009/06/04/645500.html)

[9] RIA Novosti (2007). "First Russian space flight participant will not lift off until 2009" (http://en.rian.ru/russia/20071005/82599456.html). RIA Novosti. . Retrieved 2007-10-28.

[10] Kevin Bonsor (2007). "How Space Tourism Works" (http://science.howstuffworks.com/space-tourism.htm). HowStuffWorks, Inc.. . Retrieved 2007-10-28.

[11] http://www.jsc.nasa.gov/history/oral_histories/StevensonRE/RES_5-13-99.pdf

[12] http://science.ksc.nasa.gov/shuttle/missions/61-c/mission-61-c.html NASA mission summary

[13] Spacefacts biography of Edward C. Aldridge Jr. (http://www.spacefacts.de/bios/astronauts/english/aldridge_edward.htm)

[14] "1991: Sharman becomes first Briton in space" (http://news.bbc.co.uk/onthisday/low/dates/stories/may/18/newsid_2380000/2380649.stm). *BBC News*. 1991-05-18. . Retrieved 2010-05-23.

[15] http://www.charlesinspace.com/

Soyuz (spacecraft)

For the launch vehicle of the same name see Soyuz (rocket family)

Soyuz spacecraft (TMA version)

Mission type	Used to carry personnel to and from the Salyut, Mir and International Space Station. Intended originally for Low Earth Orbit(LEO) missions.
Mission duration	Up to six months docked at a space station
Dimensions	Height 24.54 ft (7.48 m) Diameter 8.92 ft (2.72 m) Volume 254.27 cu ft (7.200 m^3)

Soyuz (Russian: Сою́з, IPA: [sɐ'jus]), *Union)* is a series of spacecraft designed for the Soviet space program by the Korolyov Design Bureau. The Soyuz succeeded the Voskhod spacecraft and was originally built as part of the Soviet Manned Lunar program.

The Soyuz spacecraft is launched by the Soyuz rocket, the most frequently used and the most reliable launch vehicle in the world.[1] [2] Soyuz rocket was initially a part of the Soyuz program, and also as part of the unmanned Zond program.

The first unmanned Soyuz mission was launched November 28, 1966; the first Soyuz mission with a crew (Soyuz 1) was launched April 23, 1967, but the cosmonaut on board, Vladimir Komarov, died during the flight's crash-landing. Soyuz 2 was an unmanned mission, and Soyuz 3, launched on October 26, 1968, was the first successful Soyuz manned mission.

Currently, the Soyuz spacecraft family is still in service.[3] Soyuz spacecraft were used to carry cosmonauts to and from Salyut and later Mir Soviet space stations, and are now used for transport to and from the International Space Station. The International Space Station maintains docked Soyuz spacecraft at all times to be used as escape craft in the event of an emergency.

Design

A Soyuz spacecraft consists of three parts (from front to back):

- A spheroid orbital module, which provides accommodation for the crew during their mission;
- A small aerodynamic reentry module, which returns the crew to Earth;
- A cylindrical service module with solar panels attached, which contains the instruments and engines.

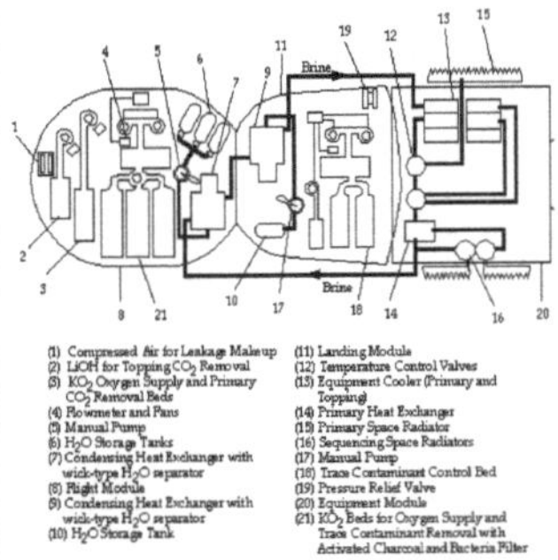

Soyuz diagram

The orbital and service modules are single-use and are destroyed upon re-entry in the atmosphere. The orbital and reentry portions are habitable living space. By moving as much equipment as possible into the orbital module, which does not have to be shielded or decelerated during atmospheric re-entry, the Soyuz is both larger and lighter than the contemporary Apollo spacecraft's command module. The Apollo command module had six cubic meters of living space and a mass of 5000 kg; the three-part Soyuz provides the same crew with nine cubic meters of living space, an airlock, and a service module for the mass of the Apollo capsule alone.

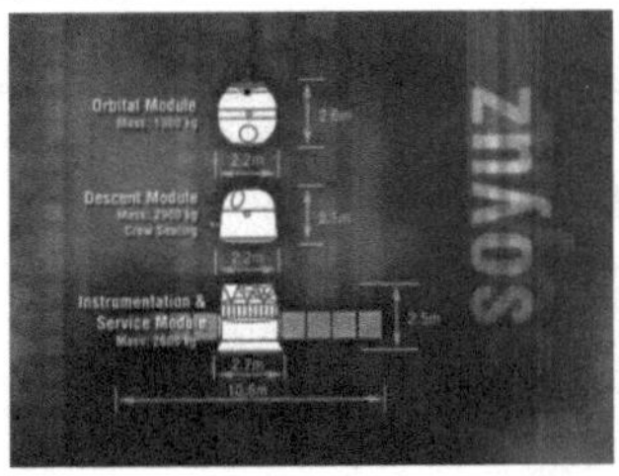

Diagram showing the three elements of the Soyuz-TMA spacecraft.

Soyuz can carry up to three crew members and provide life support for them for about 30 person days. The life support system provides a nitrogen/oxygen atmosphere at sea level partial pressures. The atmosphere is regenerated through KO_2 cylinders, which absorb most of the CO_2 and water produced by the crew and regenerates the oxygen, and LiOH cylinders which absorb leftover CO_2.

The vehicle is protected during launch by a nose fairing, which is jettisoned after passing through the atmosphere. It has an automatic docking system. The ship can be operated automatically, or by a pilot independently of ground control.

Orbital Module

Soyuz spacecraft's Orbital Module

The forepart of the spacecraft is the orbital module (Russian: *бытовой отсек (БО)*; Bytovoi otsek (BO)) also known as Habitation section. It houses all the equipment that will not be needed for reentry, such as experiments, cameras or cargo. The module also contains a toilet, docking avionics and communications gear. On the latest Soyuz versions, a small window was introduced, providing the crew with a forward view.

A hatch between it and the descent module can be closed so as to isolate it to act as an airlock if needed, cosmonauts exiting through its side port (near the descent module). On the launch pad, the astronauts enter the spacecraft through this port.

This separation also lets the orbital module be customized to the mission with less risk to the life-critical descent module. The convention of orientation in zero gravity differs from that of the descent module, as cosmonauts stand or sit with their heads to the docking port.

Reentry Module

Soyuz spacecraft's Descent Module

The reentry module (Russian: *спускаемый аппарат (СА)*; Spuskaemyi apparat (SA)) is used for launch and the journey back to Earth.Half of the reentry module is covered by a heat-resistant covering to protect it during re-entry.One half is heavier than the other,so the heavier side is facing the earth during re-entry,therefore shielding the module without having to cover the whole module. It is slowed initially by the atmosphere, then by a braking parachute, followed by the main parachute which slows the craft for landing. At one meter above the ground, solid-fuel braking engines mounted behind the heat shield are fired to give a soft landing. One of the design requirements for the reentry module was for it to have the highest possible volumetric efficiency (internal volume divided by hull area). The best shape for this is a sphere, but such a shape can provide no lift, which results in a purely ballistic reentry. Ballistic reentries are hard on the occupants due to high deceleration and cannot be steered beyond their initial deorbit burn. That is why it was decided to go with the "headlight" shape that the Soyuz uses—a hemispherical forward area joined by a barely angled conical section (seven degrees) to a classic spherical section heat shield. This shape allows a small amount of lift to be generated due to the unequal weight distribution. The nickname was thought up at a time when nearly every headlight was circular.

Service Module

Soyuz spacecraft's Instrumentation/Propulsion Module

At the back of the vehicle is the service module (Russian: *приборно-агрегатный отсек (ПАО)*; Priborno-Agregatnyi Otsek (PAO)). It has a pressurized container shaped like a bulging can (*Instrumentation compartment, PO (Priborniy Otsek*) that contains systems for temperature control, electric power supply, long-range radio communications, radio telemetry, instruments for orientation and control. A non-pressurized part of the service module (*Propulsion compartment, AO (Agregatniy Otsek)*) contains the main engine and a spare: liquid-fuel propulsion systems for maneuvering in orbit and initiating the descent back to Earth. The ship also has a system of low-thrust engines for orientation, attached to the Intermediate compartment (*PkhO or Perekhodnoi Otsek*). Outside the service module are the sensors for the orientation system and the solar array, which is oriented towards the sun by rotating the ship.

Re-entry procedure

Because its modular construction differs from that of previous designs, the Soyuz has an unusual sequence of events prior to re-entry. The spacecraft is turned engine-forward and the main engine is fired for de-orbiting fully 180° ahead of its planned landing site. This requires the least propellant for re-entry, the spacecraft traveling on an elliptical Hohmann orbit to a point where it will be low enough in the atmosphere to re-enter.

Early Soyuz spacecraft would then have the service and orbital modules detach simultaneously. As they are connected by tubing and electrical cables to the descent module, this would aid in their separation and avoid having the descent module alter its orientation. Later Soyuz spacecraft detach the orbital module before firing the main engine, which saves even more propellant, enabling the descent module to return more payload. But in no case can the orbital module remain in orbit as an addition to a space station, for the hatch enabling it to function as an airlock is part of the descent module.

Re-entry firing is done on the "dawn" side of the earth, so that the spacecraft can be seen by recovery helicopters as it descends in the evening twilight, illuminated by the sun when it is above the shadow of the Earth. The Soyuz craft is designed to come down on land, usually somewhere in the deserts of Kazakhstan in central Asia. This is in contrast to early US manned missions which "splashed down" in the ocean.

Spacecraft systems

- **Thermal Control System**—*Sistema Obespecheniya Teplovogo Rezhima, SOTR*
- **Life Support System**—*Kompleks Sredstv Obespecheniya Zhiznideyatelnosti, KSOZh*
- **Power Supply System**—*Sistema Elektropitaniya, SEP*
- **Communication and Tracking Systems**—Rassvet (Dawn) radio communications system, Onboard Measurement System (SBI), Kvant-V spacecraft control, Klyost-M television system, Orbit Radio Tracking (RKO)
- **Onboard Complex Control System**—*Sistema Upravleniya Bortovym Kompleksom, SUBK*
- **Combined Propulsion System**—*Kompleksnaya Dvigatelnaya Ustanovka, KDU*
- **Chaika-3 Motion Control System (SUD)**
- *'Optical/Visual Devices (OVP)*—VSK-4 (Vizir Spetsialniy Kosmicheskiy-4),Night Vision Device (VNUK-K, Visir Nochnogo Upravleniya po Kursu), Docking light, Pilot's Sight (VP-1, Vizir Pilota-1), Laser Range Finder (LPR-1, Lazerniy Dalnomer-1)
- **Kurs rendezvous system**
- **Docking System**—*Sistema Stykovki i Vnutrennego Perekhoda, SSVP*
- **Teleoperator Control Mode**—*Teleoperatorniy Rezhim Upravleniya, TORU*
- **Entry Actuators System**—*Sistema Ispolnitelnikh Organov Spuska, SIO-S*
- **Landing Aids Kit**—*Kompleks Sredstv Prizemleniya, KSP*
- **Portable Survival Kit**—*Nosimiy Avariyniy Zapas, NAZ,* containing a TP-82 or other pistol
- **Soyuz Launch Escape System**—*Sistema Avariynogo Spaseniya, SAS*

Variants

The Soyuz spacecraft has been the subject of continuous evolution since the early 1960s. Thus several different actual versions, proposals and projects exist.

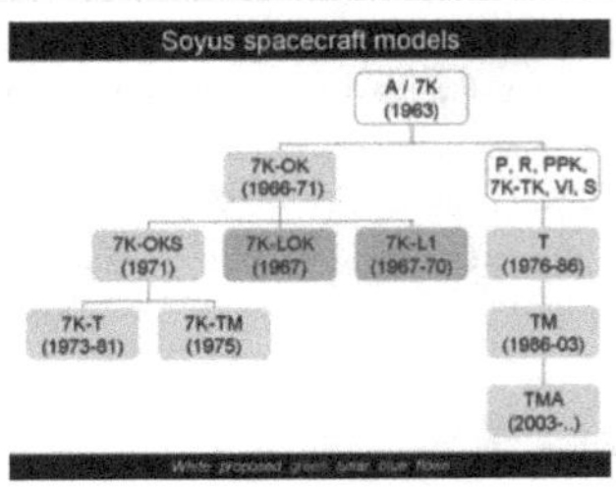

Soyuz family tree

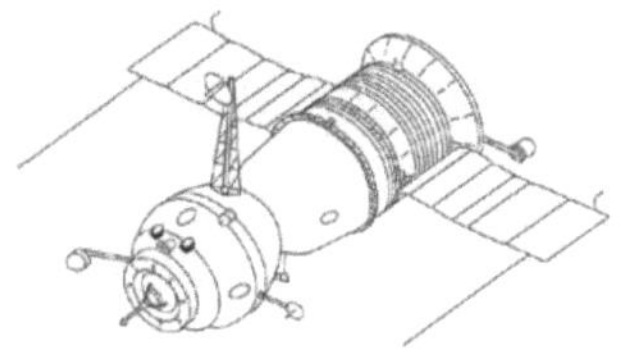

Salyut 1-type Soyuz 7K-T/A9 for three cosmonauts without space suits.

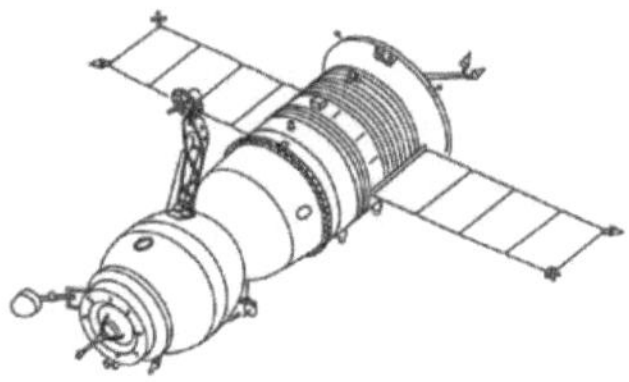

Soyuz-T spacecraft.

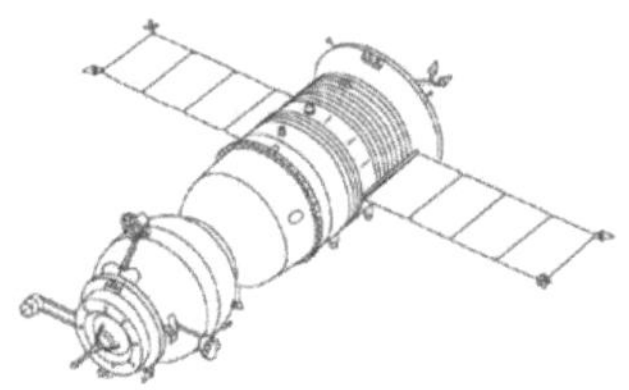

Soyuz-TM spacecraft. Compare the antennae on the orbital module to those on Soyuz-T. Differences reflect the change from the Igla rendezvous system used on Soyuz-T to the Kurs rendezvous system used on Soyuz-TM.

The Soyuz TMA-6

Technical data

Version:	Soyuz A (1963)	Soyuz 7K-OK (1967–1971)	Soyuz 7K-L3 (LOK)	Soyuz 7K-T (1973–1981)	Soyuz 7K-TM (1975)	Soyuz-T (1976–1986)	Soyuz-TM (1986–2002)	Soyuz-TMA (2003–....)
Total								
Mass (kg)	5 880	6 560	9 850	6 800	6 680	6 850	7 250	7 220
Length (m)	7.40	7.95	10.06	7.48	7.48	7.48	7.48	7.48
Max Diameter (m)	2 .50	2 .72	2 .930	2 .72	2 .72	2 .72	2 .72	2 .72
Span (m)	?	9 .80	?	9 .80/–	8 .37	10 .6	10 .6	10 .7
Orbital Module (BO)								
Mass (kg)	1 000	1 100	-	1 350	1 224	1 100	1 450	1 370
Length (m)	3 .00	3 .45	2 .26	2 .98	3 .10	2 .98	2 .98	2 .98
Diameter (m)	2 .20	2 .25	2 .3	2 .26	2 .26	2 .26	2 .26	2 .26
Volume (m³)	2 .20	5 .00	-	5 .00	5 .00	5 .00	5 .00	5 .00
Reentry module (SA)								
Mass (kg)	2 480	2 810	2 804	2 850	2 802	3 000	2 850	2 950
Length (m)	2 .30	2 .24	2 .19	2 .24	2 .24	2 .24	2 .24	2 .24
Diameter (m)	2 .17	2 .17	2 .2	2 .17	2 .17	2 .17	2 .17	2 .17
Volume (m³)	4 .00	4 .00	-	3 .50	4 .00	4 .00	3 .50	3 .50
Service module (PAO)								
Mass (kg)	2 400	2 650	-	2 700	2 654	2 750	2 950	2 900
Usable mass (kg)	830	500	3152[4]	500	500	700	880	880
Length (m)	2 .10	2 .26	2 .82	2 .26	2 .26	2 .26	2 .26	2 .26
Diameter (m)	2 .50	2 .72	2 .2	2 .72	2 .72	2 .72	2 .72	2 .72

Soyuz A (part of the "circumlunar complex") (1963)

Korolyov initially promoted the Soyuz A-B-C circumlunar complex (*7K-9K-11K*) concept (also known as L1) in which a two-man craft Soyuz-A 7K would rendezvous with other components (9K and 11K) in Earth orbit to assemble a lunar excursion vehicle, the components being delivered by the proven R-7 rocket.

First generation

The manned Soyuz spacecraft can be classified into design generations. Soyuz 1 through Soyuz 11 (1967–1971) were first-generation vehicles, carrying a crew of up to three without spacesuits and distinguished from those following by their bent solar panels and their use of the Igla automatic docking navigation system, which required special radar antennas. This first generation was called Soyuz 7K-OK and encompassed the original Soyuz and

Salyut 1 Soyuz.

The Soyuz 7K-L1 was designed to launch men from the Earth to circle the moon, and was the primary hope for a Soviet circumlunar flight. It had several test flights in the Zond program from 1967–1970 (Zond 4 to Zond 8), which produced multiple failures in the 7K-L1's re-entry systems. So the remaining 7K-L1s were scrapped. The Soyuz 7K-L3 was designed and developed in parallel to the Soyuz 7K-L1, but was also scrapped.

The next manned version of the Soyuz was the Soyuz 7K-OKS. It was designed for space station flights and had a docking port that allowed internal transfer between spacecraft. The Soyuz 7K-OKS had two manned flights, both in 1971, and the second of which depressurized upon re-entry, killing its three-man crew.

Second generation

The second generation, called *Soyuz Ferry* or Soyuz 7K-T, comprised Soyuz 12 through Soyuz 40 (1973–1981). It was developed out of the military Soyuz concepts studied in previous years. The Soyuz 7K-T/A9 version was used for the flights to the military Almaz space station.

Soyuz 7K-TM was the spacecraft used in the Apollo-Soyuz Test Project in 1975, which saw the first and only docking of a Soyuz spacecraft with an Apollo spacecraft. It was also flown in 1976 for the earth-science mission, Soyuz 22. Soyuz 7K-TM served as a technological bridge to the third generation, the Soyuz-T spacecraft.

Several military Soyuz models were planned, but none actually flew in space. These versions were named *Soyuz 7K-P*, *Soyuz 7K-PPK*, *Soyuz R*, *Soyuz 7K-VI*, and *Soyuz OIS (Orbital Research Station)*.

Third generation

The third generation Soyuz-T (T: транспортный, *Transportnyi* meaning transport) spacecraft (1976–1986) featured solar panels allowing longer missions, a revised Igla rendezvous system and new translation/attitude thruster system on the Service module. It could carry a crew of three, now wearing spacesuits.

Fourth generation

The Soyuz-TM crew transports (M: модифицированный, *Modifitsirovannyi* meaning modified) were fourth generation Soyuz spacecraft, and were used from 1986 to 2003 for ferry flights to Mir and the International Space Station.

Soyuz-TMA (2003–....)

Soyuz TMA (A: антропометрический, *Antropometricheskii* meaning anthropometric) features several changes to accommodate requirements requested by NASA in order to service the International Space Station, including more latitude in the height and weight of the crew and improved parachute systems. It is also the first expendable vehicle to feature "glass cockpit" technology. Soyuz-TMA looks identical to a Soyuz-TM spacecraft on the outside, but interior differences allow it to accommodate taller occupants with new adjustable crew couches.

Soyuz TMA-M (2010/....)

In 2004, Russian space officials announced that they intended to replace Soyuz with the new Kliper and Parom spacecrafts by early 2011.[5] Since then, Kliper appears to have been indefinitely postponed due to lack of government funding. It has since been announced that the Soyuz will receive an upgrade to make it suitable for up to one year in space, as well as new digital interior displays and updated docking equipment. This new version, known as Soyuz TMA-M, could enter service around 2010.[6]

Soyuz ACTS (2012/....)

Soyuz ACTS (Advanced Crew Transportation System), also known as Soyuz-K, is a proposed version of the Soyuz design capable of achieving lunar orbit. The upgrades could include a new habitation module developed by the European Space Agency. A novel, rocket-based precision landing system may also be implemented.[7] Missions could be launched from Baikonur or Guiana Space Centre.[8]

Related craft

The unmanned Progress spacecraft were derived from Soyuz and are used for servicing space stations.

While not being the direct derivatives of the Soyuz, the Chinese Shenzhou spacecraft and the Indian Orbital Vehicle follow the general template as pioneered by the Soyuz.

Operators

- Russian Federation: 24 August 1991 (independence of Russia from the USSR) to the present
- Union of Soviet Socialist Republics: 1966 to 26 December 1991 (Dissolution of the Soviet Union, by the dissolution of the Supreme Soviet)

See also

- Progress spacecraft
- Chinese Shenzhou spacecraft
- ISRO Orbital Vehicle
- Sokol space suit
- Space exploration
- Human spaceflight
- Crew Space Transportation System study to develop a European-Russian successor to Soyuz
- Zarya spacecraft project

Missions

See List of Soviet and Russian manned space missions as well as the Zond program

Image gallery

Early 7K-OK Soyuz at National Space Centre, Leicester, England

Soyuz spacecraft of the Apollo Soyuz Test Project (ASTP)

Soyuz 19 spacecraft as seen from Apollo CM

Soyuz docked to Mir

Soyuz spacecraft mock-up

Soyuz launch vehicles engines

Soyuz TM-31 Moves to Launch Pad, 29 October 2000

Soyuz TMA-2 launch, Baikonur, April 26, 2003

Soyuz TMA-3 landing (NASA)

Space Shuttle Orbiter and Soyuz-TM (drawn to scale).

Soyuz spacecraft on display at the Kennedy Space Center.

Soviet Insignia on an older Soyuz spacecraft.

Astronauts within Soyuz trainer.

External links

- Russia New Russian spaceship will be able to fly to Moon - space corp [9]
- RSC Energia: Concept Of Russian Manned Space Navigation Development [10]
- Mir Hardware Heritage
  - David S.F. Portree, *Mir Hardware Heritage* [11], NASA RP-1357, 1995
  - Mir Hardware Heritage (wikisource)
- Information on Soyuz spacecraft [12]
- OMWorld's ASTP Docking Trainer Page [13]
- NASA - Russian Soyuz TMA Spacecraft Details [14]
- Space Adventures circum-lunar mission - details [15]

References

[1] Soyuz launch vehicle: The most reliable means of space travel (http://www.esa.int/esaMI/Delta_Mission/SEM8XK57ESD_0.html), ESA
[2] France and Russia agree: Soyuz will launch from Kourou in French Guyana (http://eu.spaceref.com/news/viewpr.html?pid=12839), SpaceRef
[3] Soyuz rocket
[4] Lunar Orbital Spacecraft (LOK) (http://www.russianspaceweb.com/lok.html)
[5] http://www.spacedaily.com/reports/Soyuz_100_Times_More_Reliable_Than_Shuttle_999.html
[6] http://www.spacedaily.com/reports/Soyuz_100_Times_More_Reliable_Than_Shuttle_999.html
[7] (http://news.bbc.co.uk/2/hi/science/nature/8024590.stm)
[8] Advanced Crew Transportation System, ACTS (http://www.russianspaceweb.com/soyuz_acts.html)
[9] http://en.rian.ru/russia/20061018/54922208.html
[10] http://www.energia.ru/english/energia/news/news-2006/public_07-01.html
[11] http://spaceflight.nasa.gov/history/shuttle-mir/references/r-documents-mirhh.htm
[12] http://www.astronautix.com/craftfam/soyuz.htm
[13] http://www.io.com/~o_m/ssh/astp/astp.htm
[14] http://www.nasa.gov/mission_pages/station/structure/elements/soyuz/spacecraft_detail.html
[15] http://www.space.com/businesstechnology/070628_private_moonshot.html

International Space Station

International Space Station

The International Space Station on 23 May 2010 as seen from the departing Space Shuttle Atlantis during STS-132.

ISS Insignia

Station statistics

NSSDC ID:	1998-067A [1]
Call sign:	*Alpha*
Crew:	6
Launch:	1998–2011
Launch pad:	KSC LC-39, Baikonur LC-1/5 & LC-81/23
Mass:	344378 kg (759220 lb)
Length:	73 m (240 ft) from PMA-2 to *Zvezda*
Width:	108.5 m (356 ft) along truss, arrays extended
Height:	c. 20 m (c. 66 ft) nadir–zenith, arrays forward–aft
Pressurised volume:	c. 835 m^3 (c. 29,500 cu ft) (18 May 2010)

Atmospheric pressure:	101.3 kPa (29.91 inHg) (1 atm)
Perigee:	336 km altitude (181 nmi)
Apogee:	346 km altitude (189 nmi)
Orbit inclination:	51.6419 degrees
Average speed:	7,706.6 m/s (27,743.8 km/h, 17,239.2 mph)
Orbital period:	c. 91 minutes
Days in orbit:	4228 (18 June 2010)
Days occupied:	3517 (18 June 2010)
Number of orbits:	c. 66727 (18 June 2010)
Orbital decay:	2 km/month
Statistics as of 27 November 2009 (unless noted otherwise)	
References: [2] [3] [4] [5] [6] [7]	
Configuration	

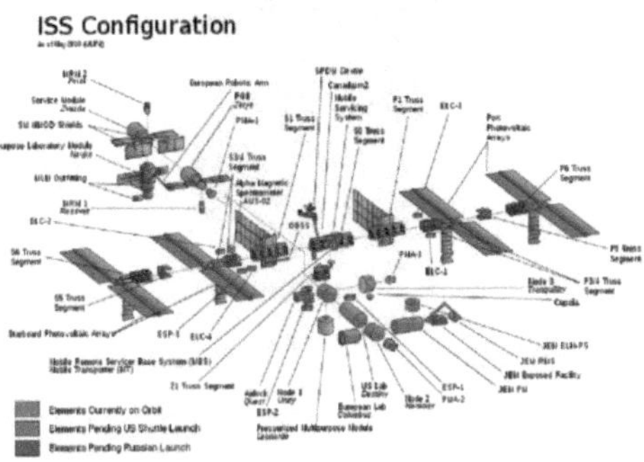

Station elements as of 18 May 2010
(exploded view)

The **International Space Station (ISS)** is an internationally developed research facility that is being assembled in low Earth orbit. On-orbit construction of the station began in 1998 and is scheduled for completion by late 2011. The station is expected to remain in operation until at least 2015, and likely 2020.[8] [9] With a greater mass than that of any previous space station, the ISS can be seen from Earth with the naked eye,[10] and is by far the largest artificial satellite that has ever orbited Earth.[11] The ISS serves as a research laboratory that has a microgravity environment in which crews conduct experiments in biology, human biology, physics, astronomy and meteorology.[12] [13] [14] The station has a unique environment for the testing of the spacecraft systems that will be required for missions to the Moon and Mars.[15] The ISS is operated by Expedition crews, with the station programme maintaining an uninterrupted human presence in space since the launch of Expedition 1 on 31 October 2000, a total of 9 years and 230 days. The programme is thus approaching the current record, set aboard *Mir*, of 3,644 days (8 days short of 10 years), with the ISS expected to take the record on 23 October 2010.[16] As of 17 March 2010, the crew of Expedition 23 is aboard.[17]

The ISS is a synthesis of several space station projects that includes the American *Freedom*, the Soviet/Russian *Mir*-2, the European *Columbus* and the Japanese *Kibō*.[18] [19] Budget constraints led to the merger of these projects into a single multi-national programme.[18] The ISS project began in 1994 with the Shuttle-*Mir* programme,[20] and the first module of the station, *Zarya*, was launched in 1998 by Russia.[18] Assembly continues, as pressurised modules, external trusses and other components are launched by American space shuttles, Russian Proton rockets and Russian Soyuz rockets.[19] As of May 2010, the station consisted of 14 pressurised modules and an extensive integrated truss structure (ITS). Power is provided by 16 solar arrays mounted on the external truss, in addition to four smaller arrays on the Russian modules.[21] The station is maintained at an orbit between 278 km (173 mi) and 460 km (286 mi) altitude, and travels at an average speed of 27,724 km/h (17,227 mph), completing 15.7 orbits per day.[22]

Operated as a joint project between the five participant space agencies, the station's sections are controlled by mission control centres on the ground operated by the American National Aeronautics and Space Administration (NASA), the European Space Agency (ESA), the Russian Federal Space Agency (RKA), the Japan Aerospace Exploration Agency (JAXA) and the Canadian Space Agency (CSA).[23] [24] The ownership and use of the space station is established in intergovernmental treaties and agreements[25] that allow the Russian Federation to retain full ownership of its own modules,[26] with the remainder of the station allocated between the other international partners.[25] The cost of the station has been estimated by ESA as €100 billion over 30 years,[27] and, although estimates range from 35 billion dollars to 160 billion dollars, the ISS is believed to be the most expensive object ever constructed.[28] The financing, research capabilities and technical design of the ISS programme have been criticised because of the high cost.[29] [30] The station is serviced by Soyuz spacecraft, Progress spacecraft, space shuttles, the Automated Transfer Vehicle and the H-II Transfer Vehicle,[24] and has been visited by astronauts and cosmonauts from 15 different nations.[11]

Purpose

The International Space Station (ISS) is an internationally developed satellite currently being assembled in Low Earth Orbit. Primarily a research laboratory, the ISS offers an advantage over spacecraft such as NASA's Space Shuttle because it is a long-term platform in the space environment, where extended studies are conducted.[11] [31] The presence of a permanent crew affords the ability to monitor, replenish, repair, and replace experiments and components of the spacecraft itself. Scientists on Earth have swift access to the crew's data and can modify experiments or launch new ones, benefits generally unavailable on specialised unmanned spacecraft.[31]

Crews, who fly expeditions of several months duration, conduct scientific experiments each day (approximately 160 man-hours a week).[12] [32] As of the conclusion of Expedition 15, 138 major science investigations had been conducted on the ISS.[33] Scientific findings, in fields from basic science to exploration research, are published every month.[15]

The ISS provides a location in the relative safety of Low Earth Orbit to test spacecraft systems that will be required for long-duration missions to the Moon and Mars. This provides experience in the maintenance, repair, and replacement of systems on-orbit, which will be essential in operating spacecraft further from Earth. Mission risks are reduced, and the capabilities of interplanetary spacecraft are advanced.[15]

Part of the crew's mission is educational outreach and international cooperation. The crew of the ISS provide opportunities for students on Earth by running student-developed experiments, making educational demonstrations, and allowing for student participation in classroom versions of ISS experiments, NASA investigator experiments, and ISS engineering activities. The ISS programme itself, with the international cooperation that it represents, allows 14 nations to live and work together in space, providing lessons for future multi-national missions.[24] [34]

Scientific research

Expedition 8 Commander and Science Officer Michael Foale conducts an inspection of the Microgravity Science Glovebox.

The ISS provides a platform to conduct experiments that require one or more of the unusual conditions present on the station. The primary fields of research include human research, space medicine, life sciences, physical sciences, astronomy and meteorology.[12] [13] [14] The 2005 NASA Authorization Act designated the American segment of the International Space Station as a national laboratory with the goal of increasing the use of the ISS by other federal agencies and the private sector.[35]

ESA astronaut Thomas Reiter, STS-116 mission specialist, works with the Passive Observatories for Experimental Microbial Systems in Micro-G (POEMS) payload in the Minus Eighty Degree Laboratory Freezer for ISS (MELFI) inside the Destiny laboratory.

Research on the ISS improves knowledge about the effects of long-term space exposure on the human body. Subjects currently under study include muscle atrophy, bone loss, and fluid shift. The data will be used to determine whether space colonisation and lengthy human spaceflight are feasible. As of 2006, data on bone loss and muscular atrophy suggest that there would be a significant risk of fractures and movement problems if astronauts landed on a planet after a lengthy interplanetary cruise (such as the six-month journey time required to fly to Mars).[36] [37] Large scale medical studies are conducted aboard the ISS via the National Space and Biomedical Research Institute (NSBRI). Prominent among these is the Advanced Diagnostic Ultrasound in Microgravity study in which astronauts (including former ISS Commanders Leroy Chiao and Gennady Padalka) perform ultrasound scans under the guidance of remote experts. The study considers the diagnosis and treatment of medical conditions in space. Usually, there is no physician onboard the ISS and diagnosis of medical conditions is a challenge. It is anticipated that remotely guided ultrasound scans will have application on Earth in emergency and rural care situations where access to a trained physician is difficult.[38] [39] [40]

Researchers are investigating the effect of the station's near-weightless environment on the evolution, development, growth and internal processes of plants and animals. In response to some of this data, NASA wants to investigate microgravity's effects on the growth of three-dimensional, human-like tissues, and the unusual protein crystals that can be formed in space.[13]

The investigation of the physics of fluids in microgravity will allow researchers to model the behaviour of fluids better. Because fluids can be almost completely combined in microgravity, physicists investigate fluids that do not mix well on Earth. In addition, an examination of reactions that are slowed by low gravity and temperatures will give scientists a deeper understanding of superconductivity.[13]

The study of materials science is an important ISS research activity, with the objective of reaping economic benefits through the improvement of techniques used on the ground.[41] Other areas of interest include the effect of the low gravity environment on combustion, through the study of the efficiency of burning and control of emissions and pollutants. These findings may improve our knowledge about energy production, and lead to economic and environmental benefits. Future plans are for the researchers aboard the ISS to examine aerosols, ozone, water vapour, and oxides in Earth's atmosphere, as well as cosmic rays, cosmic dust, antimatter, and dark matter in the universe.[13]

Origins

Space Shuttle *Atlantis* docked to *Mir* on STS-71, during the Shuttle-Mir Program

The International Space Station represents a union of several national space station projects that originated during the Cold War. In the early 1980s, NASA planned to launch a modular space station called *Freedom* as a counterpart to the Soviet *Salyut* and *Mir* space stations, while the Soviets were planning to construct *Mir*-2 in the 1990s as a replacement for *Mir*.[18] Because of budget and design constraints, *Freedom* never progressed past mock-ups and minor component tests.

With the fall of the Soviet Union and the end of the Space Race, *Freedom* was nearly cancelled by the United States House of Representatives. The post-Soviet economic chaos in Russia led to the cancellation of *Mir*-2, though only after its base block, DOS-8, had been constructed.[18] Similar budgetary difficulties were faced by other nations with space station projects, which prompted the American government to negotiate with European states, Russia, Japan, and Canada in the early 1990s to begin a collaborative project.[18]

In June 1992 American president George H. W. Bush and Russian president Boris Yeltsin agreed to cooperate on space exploration. The resulting *Agreement between the United States of America and the Russian Federation Concerning Cooperation in the Exploration and Use of Outer Space for Peaceful Purposes* called for a short, joint space programme, with one American astronaut deployed to the Russian space station *Mir* and two Russian cosmonauts deployed to a Space Shuttle.[18]

In September 1993, American Vice-President Al Gore, Jr., and Russian Prime Minister Viktor Chernomyrdin announced plans for a new space station, which eventually became the International Space Station.[42] They also agreed, in preparation for this new project, that the United States would be heavily involved in the *Mir* programme as part of an agreement that later included Space Shuttle orbiters docking with *Mir*.[20]

According to the plan, the International Space Station programme would combine the proposed space stations of all participant agencies: NASA's *Freedom*, the RSA's *Mir*-2 (with DOS-8 later becoming *Zvezda*), ESA's *Columbus*, and the Japanese *Kibō* laboratory. When the first module, *Zarya*, was launched in 1998, the station was expected to be completed by 2003. Delays have led to a revised estimated completion date of 2011.[43]

Station structure

Assembly

The assembly of the International Space Station, a major endeavour in space architecture, began in November 1998.[3] Astronauts install each element using spacewalks. By 27 November 2009, they had completed 136, totalling 849 hours of extra-vehicular activity (EVA), all devoted to assembly and maintenance of the station. Twenty-eight of these spacewalks originated from the airlocks of docked Space Shuttles; the remaining 108 were launched from the station.[2]

Astronaut Ron Garan during an STS-124 ISS assembly spacewalk

The first segment of the ISS, *Zarya*, was launched on 20 November 1998 on a Russian Proton rocket, followed two weeks later by *Unity*—the first of three node modules—which was launched aboard Space Shuttle flight STS-88. This bare two-module core of the ISS remained unmanned for the next one-and-a-half years. In July 2000 the Russian module *Zvezda* was added, allowing a maximum crew of three to occupy the ISS continuously. The first resident crew, Expedition 1, arrived in November 2000 on Soyuz TM-31, midway between the flights of STS-92 and STS-97. These two Space Shuttle flights each added segments of the station's Integrated Truss Structure, which provided the embryonic station with communications, guidance, electrical grounding (on Z1), and power via solar arrays located on the P6 truss.[44]

Over the next two years the station continued to expand. A Soyuz-U rocket delivered the *Pirs* docking compartment. The Space Shuttles *Discovery*, *Atlantis*, and *Endeavour* delivered the *Destiny* laboratory and *Quest* airlock, in addition to the station's main robot arm, the *Canadarm2*, and several more segments of the Integrated Truss Structure.[44]

The expansion schedule was interrupted by the destruction of the Space Shuttle *Columbia* on STS-107 in 2003, with the resulting hiatus in the Space Shuttle programme halting station assembly until the launch of *Discovery* on STS-114 in 2005.[45]

The official resumption of assembly was marked by the arrival of *Atlantis*, flying STS-115, which delivered the station's second set of solar arrays. Several more truss segments and a third set of arrays were delivered on STS-116, STS-117, and STS-118. As a result of the major expansion of the station's power-generating capabilities, more pressurised modules could be accommodated, and the *Harmony* node and *Columbus* European laboratory were added. These were followed shortly after by the first two components of *Kibō*, the Japanese Experiment Module. In March 2009, STS-119 completed the Integrated Truss Structure with the installation of the fourth and final set of solar arrays. The final section of *Kibō* was delivered in July 2009 on STS-127, followed by the Russian *Poisk* module. The third node, *Tranquility*, was delivered in February 2010 during STS-130 by the Space Shuttle *Endeavour*, alongside the Cupola, closely followed in May 2010 by the penultimate Russian module, 'Rassvet, delivered by Space Shuttle *Atlantis* on STS-132.

As of May 2010, the station consisted of fourteen pressurised modules and the complete Integrated Truss Structure. Still to be launched is the Pressurised Multipurpose Module *Leonardo*, the Russian Multipurpose Laboratory Module *Nauka* and a number of external components, including the European Robotic Arm and Alpha Magnetic Spectrometer (AMS-02). Assembly is expected to be completed by 2011, by which point the station will have a mass in excess of 400 metric tons (440 short tons).[3] [43]

Pressurised modules

When completed, the ISS will consist of sixteen pressurised modules with a combined volume of around 1000 cubic metres (35000 cu ft). These modules include laboratories, docking compartments, airlocks, nodes and living quarters. Thirteen of these components are already in orbit, with the remaining three awaiting launch. Each module was or will be launched either by the Space Shuttle, Proton rocket or Soyuz rocket.[44]

<table>
<tr><th>Module</th><th>Assembly mission</th><th>Launch date</th><th>Launch system</th><th>Nation</th><th>Isolated View</th><th></th></tr>
<tr><td rowspan="2">Zarya
(lit. 'dawn')
(FGB)</td><td>1A/R</td><td>20 November 1998</td><td>Proton-K</td><td>Russia (Builder)
USA (Financier)</td><td rowspan="2"></td><td rowspan="2">[46]</td></tr>
<tr><td colspan="4">The first component of the ISS to be launched, Zarya provided electrical power, storage, propulsion, and guidance during initial assembly. The module now serves as a storage compartment, both inside the pressurised section and in the externally mounted fuel tanks.</td></tr>
<tr><td rowspan="2">Unity
(Node 1)</td><td>2A</td><td>4 December 1998</td><td>Space Shuttle Endeavour, STS-88</td><td>USA</td><td rowspan="2"></td><td rowspan="2">[47]</td></tr>
<tr><td colspan="4">The first node module, connecting the American section of the station to the Russian section (via PMA-1), and providing berthing locations for the Z1 truss, Quest airlock, Destiny laboratory and Tranquillity node.</td></tr>
<tr><td rowspan="2">Zvezda
(lit. 'star')
(Service Module)</td><td>1R</td><td>12 July 2000</td><td>Proton-K</td><td>Russia</td><td rowspan="2">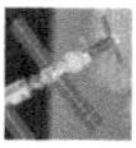</td><td rowspan="2">[48]</td></tr>
<tr><td colspan="4">The station's service module, which provides the main living quarters for resident crews, environmental systems and attitude & orbit control. The module also provides docking locations for Soyuz spacecraft, Progress spacecraft and the Automated Transfer Vehicle, and its addition rendered the ISS permanently habitable for the first time.</td></tr>
<tr><td rowspan="2">Destiny
(US Laboratory)</td><td>5A</td><td>7 February 2001</td><td>Space Shuttle Atlantis, STS-98</td><td>USA</td><td rowspan="2">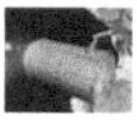</td><td rowspan="2">[49]</td></tr>
<tr><td colspan="4">The primary research facility for US payloads aboard the ISS, Destiny is intended for general experiments. The module houses 24 International Standard Payload Racks, some of which are used for environmental systems and crew daily living equipment, and features a 51-centimetre (20 in) optically perfect window, the largest such window ever produced for use in space. Destiny also serves as the mounting point for most of the station's Integrated Truss Structure.</td></tr>
<tr><td rowspan="2">Quest
(Joint Airlock)</td><td>7A</td><td>12 July 2001</td><td>Space Shuttle Atlantis, STS-104</td><td>USA</td><td rowspan="2"></td><td rowspan="2">[50]</td></tr>
<tr><td colspan="4">The primary airlock for the ISS, Quest hosts spacewalks with both US EMU and Russian Orlan spacesuits. Quest consists of two segments; the equipment lock, that stores spacesuits and equipment, and the crew lock, from which astronauts can exit into space.</td></tr>
<tr><td rowspan="2">Pirs
(lit. 'pier')
(Docking Compartment)</td><td>4R</td><td>14 September 2001</td><td>Soyuz-U, Progress M-SO1</td><td>Russia</td><td rowspan="2">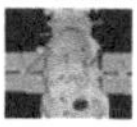</td><td rowspan="2">[51]</td></tr>
<tr><td colspan="4">Pirs provides the ISS with additional docking ports for Soyuz and Progress spacecraft, and allows egress and ingress for spacewalks by cosmonauts using Russian Orlan spacesuits, in addition to providing storage space for these spacesuits.</td></tr>
<tr><td rowspan="2">Harmony
(Node 2)</td><td>10A</td><td>23 October 2007</td><td>Space Shuttle Discovery, STS-120</td><td>Europe (Builder)
USA (Operator)</td><td rowspan="2">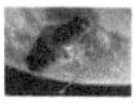</td><td rowspan="2">[52]</td></tr>
<tr><td colspan="4">The second of the station's node modules, Harmony is the utility hub of the ISS. The module contains four racks that provide electrical power, bus electronic data, and acts as a central connecting point for several other components via its six Common Berthing Mechanisms (CBMs). The European Columbus and Japanese Kibō laboratories are permanently berthed to the module, and American Space Shuttle Orbiters dock with the ISS via PMA-2, attached to Harmony's forward port. In addition, the module serves as a berthing port for the Italian Multi-Purpose Logistics Modules during shuttle logistics flights.</td></tr>
</table>

Columbus (European Laboratory)	1E	7 February 2008	Space Shuttle *Atlantis*, STS-122	Europe	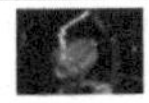	[53] [54]
	The primary research facility for European payloads aboard the ISS, *Columbus* provides a generic laboratory as well as facilities specifically designed for biology, biomedical research and fluid physics. Several mounting locations are affixed to the exterior of the module, which provide power and data to external experiments such as the European Technology Exposure Facility (EuTEF), Solar Monitoring Observatory, Materials International Space Station Experiment, and Atomic Clock Ensemble in Space. A number of expansions are planned for the module to study quantum physics and cosmology.					
Kibō Experiment Logistics Module (lit. 'hope' and 'wish' JEM–ELM)	1J/A	11 March 2008	Space Shuttle *Endeavour*, STS-123	Japan	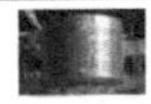	[55]
	Part of the *Kibō* Japanese Experiment Module laboratory, the ELM provides storage and transportation facilities to the laboratory with a pressurised section to serve internal payloads.					
Kibō Pressurised Module (JEM–PM)	1J	31 May 2008	Space Shuttle *Discovery*, STS-124	Japan	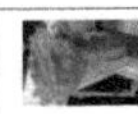	[55] [56]
	Part of the *Kibō* Japanese Experiment Module laboratory, the PM is the core module of *Kibō* to which the ELM and Exposed Facility are berthed. The laboratory is the largest single ISS module and contains a total of 23 racks, including 10 experiment racks. The module is used to carry out research in space medicine, biology, Earth observations, materials production, biotechnology, and communications research. The PM also serves as the mounting location for an external platform, the Exposed Facility (EF), that allows payloads to be directly exposed to the harsh space environment. The EF is serviced by the module's own robotic arm, the JEM–RMS, which is mounted on the PM.					
Poisk (lit. 'search') (Mini-Research Module 2)	5R	10 November 2009	Soyuz-U, Progress M-MIM2	Russia		[57] [58]
	One of the Russian ISS components, MRM2 will be used for docking of Soyuz and Progress ships, as an airlock for spacewalks and as an interface for scientific experiments.					
Tranquility (Node 3)	20A	8 February 2010	Space Shuttle *Endeavour*, STS-130	Europe (Builder) USA (Operator)		[59] [60]
	The third and last of the station's US nodes, *Tranquility* contains an advanced life support system to recycle waste water for crew use and generate oxygen for the crew to breathe. The node also provides four berthing locations for more attached pressurised modules or crew transportation vehicles, in addition to the permanent berthing location for the station's Cupola.					
Cupola	20A	8 February 2010	Space Shuttle *Endeavour*, STS-130	Europe (Builder) USA (Operator)	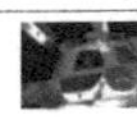	[61]
	The Cupola is an observatory module that provides ISS crew members with a direct view of robotic operations and docked spacecraft, as well as an observation point for watching the Earth. The module comes equipped with robotic workstations for operating the SSRMS and shutters to protect its windows from damage caused by micrometeorites.					
Rassvet (lit. 'dawn') (Mini-Research Module 1)	ULF4	14 May 2010	Space Shuttle *Atlantis*, STS-132	Russia		[43]
	MRM1 is being used for docking and cargo storage aboard the station.					

Scheduled to be launched

Module	Assembly mission	Launch date	Launch system	Nation	Isolated View	
Leonardo (Pressurised Multipurpose Module)	ULF5	NET 16 September 2010	Space Shuttle *Discovery*, STS-133	Italy (Builder) USA (Operator)		[62] [63] [64]
	The *Leonardo* PMM will house spare parts and supplies, allowing longer times between resupply missions and freeing space in other modules, particularly *Columbus*. The PMM was created by converting the Italian *Leonardo* Multi-Purpose Logistics Module into a module that could be permanently attached to the station. The arrival of the module will mark the completion of the US Orbital Segment.					
Nauka (lit. 'science') (Multipurpose Laboratory Module)	3R	c. December 2011	Proton-M	Russia		[43] [65]
	The MLM will be Russia's primary research module as part of the ISS and will be used for general microgravity experiments, docking, and cargo logistics. The module provides a crew work and rest area, and will be equipped with a backup attitude control system that can be used to control the station's attitude. Based on the current assembly schedule, the arrival of *Nauka* will complete construction of the Russian Orbital Segment and it will be the last major component added to the station.					

Cancelled modules

Several modules planned for the station have been cancelled over the course of the ISS programme, whether for budgetary reasons, because the modules becoming unnecessary, or following a redesign of the station after the 2003 *Columbia* disaster. The cancelled modules include:

The prototype X-38 lifting body, the cancelled ISS Crew Return Vehicle

- The US Centrifuge Accommodations Module for experiments in varying levels of artificial gravity.[66]
- The US Habitation Module, which would have served as the station's living quarters. The sleep stations are now spread throughout the station.[67]
- The US Crew Return Vehicle would have served as the station's lifeboat; a service now provided by one Soyuz spacecraft for every three crew members aboard.[68]
- The US Interim Control Module and ISS Propulsion Module were intended to replace functions of *Zvezda* in case of a launch failure.[69]
- The Russian Universal Docking Module, to which the cancelled Russian Research modules and spacecraft would have docked.[70]
- The Russian Science Power Platform would have provided the Russian Orbital Segment with a power supply independent of the ITS solar arrays.[70]
- Two Russian Research Modules that were planned to be used for scientific research.[71]

Unpressurised elements

Astronaut Stephen K. Robinson anchored to the end of Canadarm2 during STS-114

In addition to the pressurised modules, the ISS features a large number of external components. The largest component is the Integrated Truss Structure (ITS), to which the station's main solar arrays and thermal radiators are mounted.[21] The ITS consists of ten separate segments forming a structure 108.5 m (356 ft) long.[3]

The Alpha Magnetic Spectrometer (AMS), a particle physics experiment, is scheduled to be launched on STS-134 in 2010, and will be mounted externally on the ITS. The AMS will measure cosmic rays and look for evidence of dark matter and antimatter.[72]

The ITS serves as a base for the main remote manipulator system called the Mobile Servicing System (MSS). This consists of the Mobile Base System (MBS), the Canadarm2, and the Special Purpose Dexterous Manipulator. The MBS rolls along rails built into some of the ITS segments to allow the arm to reach all parts of the US segment of the station.[73] The MSS is due to have its reach increased by an Orbiter Boom Sensor System, scheduled for installation during the STS-133 mission. [74]

Two other remote manipulator systems are present in the station's final configuration. The European Robotic Arm, which will service the Russian Orbital Segment, will be launched alongside the Multipurpose Laboratory Module. [75] The JEM RMS, which services the JEM Exposed Facility,[76] was launched on STS-124 and is attached to the JEM Pressurised Module. In addition to these robotic arms, there are two Russian *Strela* cargo cranes used for moving spacewalking cosmonauts and parts around the exterior of the Russian Orbital Segment.[77]

The station in its complete form will have several smaller external components, such as the three External Stowage Platforms (ESPs), launched on STS-102, STS-114 and STS-118, which are used to store spare parts. Four ExPRESS Logistics Carriers (ELCs) will allow experiments to be deployed and conducted in the vacuum of space, and will provide the necessary electricity and computing to process experimental data locally. ELCs 1 and 2 were delivered on STS-129 in November 2009, and ELCs 3 and 4 are scheduled for delivery on STS-134 in November 2010 and STS-133 in September 2010.[43] [78] There are two exposure facilities mounted directly to laboratory modules: the JEM Exposed Facility serves as an external 'porch' for the Japanese Experiment Module complex,[79] and a facility on the European *Columbus* laboratory provides power and data connections for experiments such as the European Technology Exposure Facility[80] [81] and the Atomic Clock Ensemble in Space.[82]

Power supply

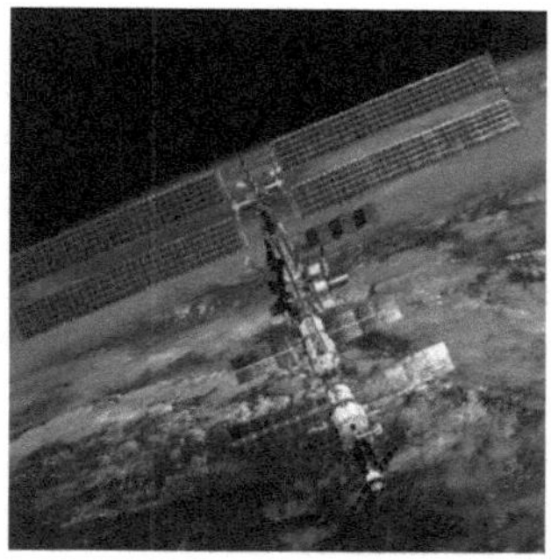

The ISS in 2001, showing the solar arrays on *Zarya* and *Zvezda*, in addition to the US P6 solar arrays

Photovoltaic (PV) arrays power the ISS. The Russian segment of the station, like the space shuttle and most aircraft, uses 28 volt DC partly provided by four solar arrays mounted directly to *Zarya* and *Zvezda*. The rest of the station uses 130-180 V DC from the US PV array arranged as four wing pairs. Each wing produces nearly 32.8 kW.[21]

Power is stabilised and distributed at 160 V DC and converted to the user-required 124 V DC. The higher distribution voltage allows smaller, lighter conductors. The two station segments share power with converters, essential since the cancellation of the Russian Science Power Platform made the Russian Orbital Segment dependent on the US arrays.[83]

The station uses rechargeable nickel-hydrogen batteries for continuous power during the 35 minutes of every 90 minute orbit that it is eclipsed by the Earth. The batteries are recharged on the day side of the earth. They have a 6.5 year lifetime (over 37,000 charge/discharge cycles) and will be regularly replaced over the anticipated 20-year life of the station.[84]

The US solar arrays normally track the sun to maximise power generation. Each array is about 375 m^2 (450 yd^2) in area and 58 metres (63 yd) long. In the complete configuration, the solar arrays track the sun by rotating the *alpha gimbal* once per orbit while the *beta gimbal* follows slower changes in the angle of the sun to the orbital plane. The Night Glider mode aligns the solar arrays parallel to the velocity vector at night to reduce the significant aerodynamic drag at the station's relatively low orbital altitude.[85]

Orbit control

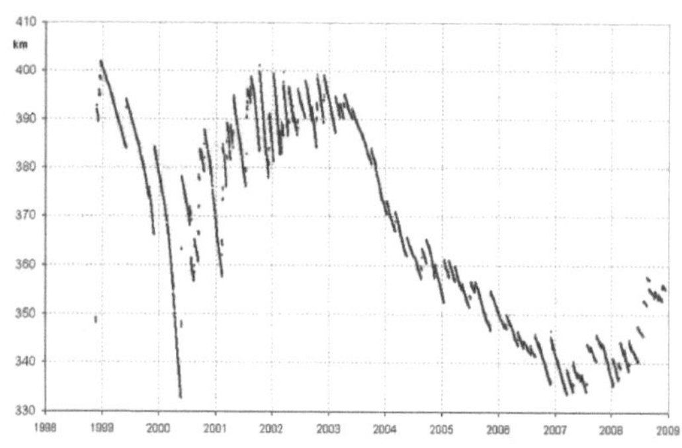

Graph showing the changing altitude of the ISS from November 1998 until January 2009

The ISS is maintained in a near circular orbit with a minimum mean altitude of 278 km (173 mi) and a maximum of 460 km (286 mi). It travels at an average speed of 27,724 kilometres (17,227 mi) per hour, and completes 15.7 orbits per day.[22] The normal maximum altitude is 425 km (264 mi) to allow Soyuz rendezvous missions. As the ISS constantly loses altitude because of a slight atmospheric drag, it needs to be boosted to a higher altitude several times each year.[31] [86] This boost can be performed by the station's two main engines on the *Zvezda* service module, a docked space shuttle, a Progress resupply vessel, or by ESA's ATV. It takes approximately two orbits (three hours) for the boost to a higher altitude to be completed.[86]

In December 2008 NASA signed an agreement with the Ad Astra Rocket Company which may result in the testing on the ISS of a VASIMR plasma propulsion engine.[87] This technology could allow station-keeping to be done more economically than at present.[88] [89] The station's navigational position and velocity, or state vector, is independently established using the US Global Positioning System (GPS) and a combination of state vector updates from Russian

Ground Sites and the Russian GLONASS system.

The attitude (orientation) of the station is independently determined by a set of sun, star and horizon sensors on *Zvezda* and the US GPS with antennas on the S0 truss and a receiver processor in the US lab. The attitude knowledge is propagated between updates by rate sensors.[24] Attitude control is maintained by either of two mechanisms; normally, a system of four control moment gyroscopes (CMGs) keeps the station oriented, with *Destiny* forward of *Unity*, the P truss on the port side, and *Pirs* on the Earth-facing (nadir) side. When the CMG system becomes 'saturated'—when the set of CMGs exceed their operational range or cannot track a series of rapid movements—they can lose their ability to control station attitude.[90] In this event, the Russian attitude control system is designed to provide desaturating thruster firings, or take over in an automated fashion. while under Russian thruster control, thrusters are used to maintain station attitude while the CMG system can be reset for control. This automatic attitude control safing has only occurred once, during Expedition 10.[91] When a space shuttle is docked to the station, it can also be used to maintain station attitude. This occurs during portions of every mated shuttle ISS mission. Shuttle control was used exclusively during STS-117 as the S3/S4 truss was installed.[92]

Communications

Radio communications provide telemetry and scientific data links between the station and Mission Control Centres. Radio links are also used during rendezvous and docking procedures and for audio and video communication between crewmembers, flight controllers and family members. As a result, the ISS is equipped with internal and external communication systems used for different purposes.[93]

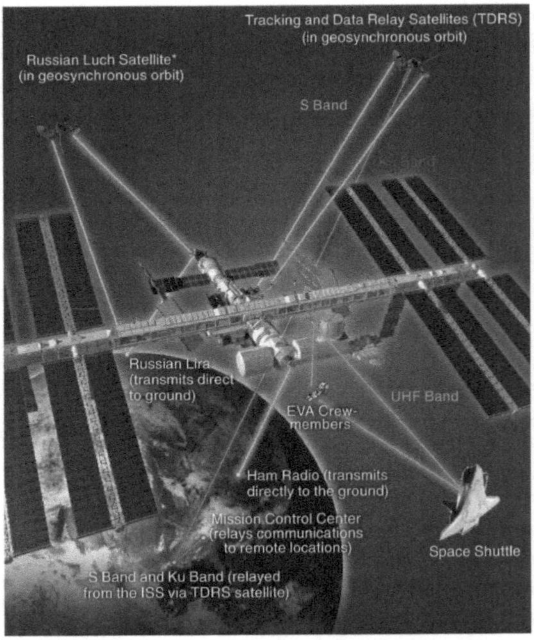

The communications systems used by the ISS

* Luch satellite not currently in use.

The Russian Orbital Segment communicates directly with the ground via the *Lira* antenna mounted to *Zvezda*.[24] [94] The *Lira* antenna also has the capability to use the *Luch* data relay satellite system.[24] This system, used for communications with *Mir*, fell into disrepair during the 1990s, and as a result is no longer in use,[18] [24] [95] although two new *Luch* satellites—*Luch*-5A and *Luch*-5B—are planned for launch in 2011 to restore the operational capability of the system.[96] Another Russian communications system is the Voskhod-M.

The US Orbital Segment (USOS) makes use of two separate radio links mounted in the Z1 truss structure: the S band (used for audio) and Ku band (used for audio, video and data) systems. These transmissions are routed via the US Tracking and Data Relay Satellite System (TDRSS) in geostationary orbit, which allows for almost continuous real-time communications with NASA's Mission Control Centre (MCC-H) in Houston.[19] [24] [93] Data channels for the Canadarm2, European *Columbus* laboratory and Japanese *Kibō* modules are routed via the S band and Ku band systems, although the European Data Relay Satellite System and a similar Japanese system will eventually complement the TDRSS in this role.[19] [97] Communications between modules are carried on an internal digital wireless network.[98]

UHF radio is used by astronauts and cosmonauts conducting EVAs. UHF is employed by other spacecraft that dock to or undock from the station, such as Soyuz, Progress, HTV, ATV and the Space Shuttle (except the shuttle also makes use of the S band and Ku band systems via TDRSS), to receive commands from Mission Control and ISS crewmembers.[24] Automated spacecraft are fitted with their own communications equipment; the ATV uses a laser attached to the spacecraft and equipment attached to *Zvezda*, known as the Proximity Communications Equipment, to accurately dock to the station.[99] [100]

Microgravity

At the station's orbital altitude, the gravity from the Earth is 88% of that at sea level. The state of weightlessness is caused by the constant free fall of the ISS, which, because of the equivalence principle, is indiscernible from a state of zero gravity.[101] The environment on the station is, however, often described as microgravity, as the weightlessness is imperfect. This is caused by four separate effects:[102]

- The drag resulting from the residual atmosphere.
- Vibratory acceleration caused by mechanical systems and the crew on board the ISS.
- Orbital corrections by the on-board gyroscopes or thrusters.
- The spatial separation from the real centre of mass of the ISS. Any part of the ISS not at the exact centre of mass will tend to follow its own orbit. However, as each point is physically part of the station, this is impossible, and so each component is subject to small accelerations from the forces which keep them attached to the station as it orbits.[102] This is also called the tidal force.

Life support

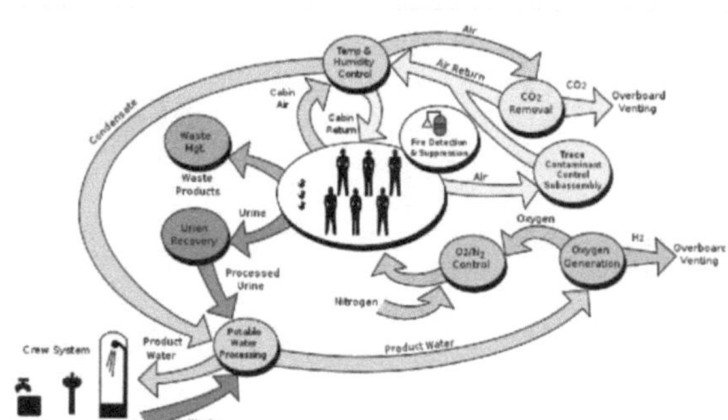

The interactions between the components of the ISS Environmental Control and Life Support System (ECLSS)

The ISS Environmental Control and Life Support System (ECLSS) provides or controls atmospheric pressure, fire detection and suppression, oxygen levels, waste management and water supply. The highest priority for the ECLSS is the ISS atmosphere, but the system also collects, processes, and stores waste and water produced and used by the crew—a process that recycles fluid from the sink, shower, toilet, and condensation from the air. The *Elektron* system aboard *Zvezda* and a similar system in *Destiny* generate oxygen aboard the station.[103] The crew has a backup option in the form of bottled oxygen and Solid Fuel Oxygen Generation (SFOG) canisters.[104] Carbon dioxide is removed from the air by the Vozdukh system in *Zvezda*. Other by-products of human metabolism, such as methane from the intestines and ammonia from sweat, are removed by activated charcoal filters.[104]

The atmosphere on board the ISS is similar to the Earth's.[105] Normal air pressure on the ISS is 101.3 kPa (14.7 psi);[106] the same as at sea level on Earth. An Earth-like atmosphere offers benefits for crew comfort, and is much safer than the alternative, a pure oxygen atmosphere, because of the increased risk of a fire such as that responsible for the deaths of the Apollo 1 crew.[107]

Sightings

A July 2007 sighting of the International Space Station.

Because of the size of the ISS (about that of an American football field) and the large reflective area offered by its solar panels, ground based observation of the station is possible with the naked eye if the observer is in the right location at the right time. In many cases, the station is one of the brightest naked-eye objects in the sky, although it is visible only for periods ranging from two to five minutes.[10]

To view the station, the following conditions need to be fulfilled, assuming the weather is clear: The station must be above the observer's horizon, and it must pass within about 2000 kilometres (1200 mi) of the observation site (the closer the better). It must be dark enough at the observer's location for stars to be visible, and the station must be in sunlight rather than in the Earth's shadow. It is common for the third condition to begin or end during what would otherwise be a good viewing opportunity. In the evening, as the station moves further from the dusk, going from west to east it will appear to suddenly fade and disappear. In the reverse situation, it may suddenly appear in the sky as it approaches the dawn.[10] [108]

The station has now become bright enough to be seen during the day under certain conditions.[109]

Politics, utilisation and financing

Legal aspects

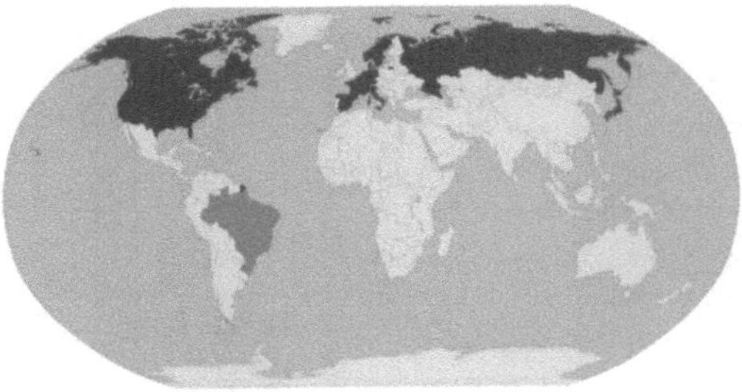

Primary contributing nations   NASA contracted nations

The ISS is a joint project of several space agencies: the US National Aeronautics and Space Administration (NASA), the Russian Federal Space Agency (RKA), the Japan Aerospace Exploration Agency (JAXA), the Canadian Space Agency (CSA) and the European Space Agency (ESA).[23]

As a multinational project, the legal and financial aspects are complex. Issues of concern include the ownership of modules, station utilisation by participant nations, and responsibilities for station resupply. Obligations and rights are established by the Space Station Intergovernmental Agreement (IGA). This international treaty was signed on 28 January 1998 by the primary nations involved in the Space Station project; the United States of America, Russian Federation, Japan, Canada and ten member states of the European Space Agency (Belgium, Denmark, France, Germany, Italy, The Netherlands, Norway, Spain, Sweden and Switzerland).[25] A second layer of agreements was then achieved, called Memoranda of Understanding (MOU), between NASA and ESA, CSA, RKA and JAXA. These agreements are then further split, such as for the contractual obligations between nations, and trading of partners' rights and obligations.[25] Use of the Russian Orbital Segment is also negotiated at this level.[26]

In addition to these main intergovernmental agreements, Brazil joins as a bilateral partner of the United States by a contract with NASA to supply hardware[110] . In return, NASA will provide Brazil with access to its ISS facilities on-orbit, as well as a flight opportunity for one Brazilian astronaut during the course of the ISS programme. Italy has a similar contract with NASA to provide comparable services, although Italy also takes part in the programme directly via its membership in ESA.[111] China has reportedly expressed interest in the project, especially if it would be able to work with the RKA. However, as of 2009 China is not involved because of US objections.[112] [113] The heads of both the South Korean and Indian space agencies announced at the first plenary session of the 2009 International Astronautical Congress on 12 October that their nations intend to join the ISS program. The talks are due to begin in 2010. The heads of agency also expressed support for extending ISS lifetime.[114]

Utilisation rights

The Russian part of the station is operated and controlled by the Russian Federation's space agency and provides Russia with the right to nearly one-half of the crew time for the ISS. The allocation of remaining crew time (three to four crew members of the total permanent crew of six) and hardware within the other sections of the station has been assigned as follows:

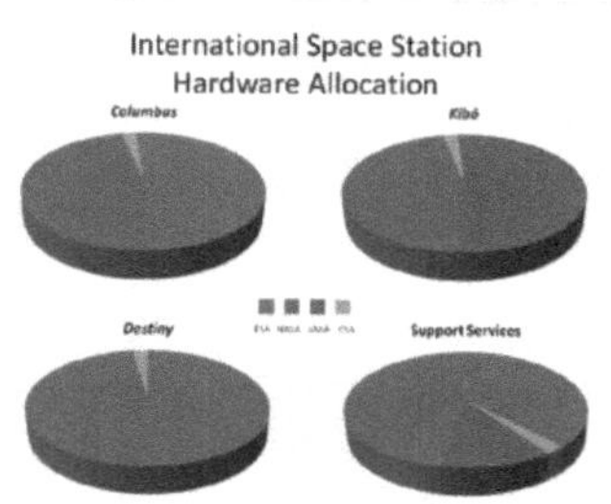

Allocation of American segment hardware utilisation between nations

- *Columbus*: 51% for the ESA, 46.7% for NASA, and 2.3% for CSA.[25]
- *Kibō*: 51% for the JAXA, 46.7% for NASA, and 2.3% for CSA.[97]
- *Destiny*: 97.7% for NASA and 2.3% for CSA.[115]
- Crew time, electrical power and rights to purchase supporting services (such as data upload and download and communications) are divided 76.6% for NASA, 12.8% for JAXA, 8.3% for ESA, and 2.3% for CSA.[25] [97] [115]

Costs

The cost estimates for the ISS range from 35 billion to 160 billion dollars.[28] ESA, the one agency which actually presents potential overall costs, estimates €100 billion for the entire station over 30 years.[27] A precise cost estimate for the ISS is unclear, as it is difficult to determine which costs should be attributed to the ISS programme, or how the Russian contribution should be measured.[28]

Criticism

Critics of the ISS contend that the time and money spent on the ISS could be better spent on other projects—whether they be robotic spacecraft missions, space exploration, investigations of problems on Earth, or just tax savings.[29] [30] Some critics, such as Robert L. Park, argue that little scientific research was convincingly planned for the ISS, and that the primary feature of a space-based laboratory, its microgravity environment, can be studied less expensively with a "vomit comet".[29] [116] [117]

The research capabilities of the ISS have been criticised, particularly following the cancellation of the ambitious Centrifuge Accommodations Module, which, alongside other equipment cancellations, means research done on the ISS is generally limited to experiments which do not require any specialised apparatus. For example, in the first half of 2007, ISS research dealt primarily with human biological responses to living and working in space, covering topics like kidney stones, circadian rhythm, and the effects of cosmic rays on the nervous system.[118] [119] [120] Other criticisms hinge on the technical design of the ISS, including the high inclination of the station's orbit, which leads to a higher cost for US-based launches to the station.[121]

In response to some of these comments, advocates of manned space exploration say that criticism of the ISS project is short-sighted, and that manned space research and exploration have produced billions of dollars' worth of benefits. NASA has estimated that the indirect economic return from spin-offs of human space exploration has been many times the initial public investment, although this estimate is based on the Apollo Program and was made in the 1970s.[122] A review of the claims by the Federation of American Scientists argued that NASA's rate of return from spin-offs is actually very low, except for aeronautics work that has led to aircraft sales.[123]

End of mission and deorbit plans

NASA planned to deorbit the ISS in the first quarter of 2016.[124] However, the plan to end the ISS programme in 2015, as determined in 2004 by then-President George W. Bush, has been rejected by the current Obama administration. With the new budget announced on 1 February 2010, the administration aims to extend the lifetime through 2020. Current NASA administrator Charles Bolden has stated the lifetime of the ISS will be extended beyond 2016, the year that international funding for the station is planned to end.[125] The Augustine Commission, which reviewed NASA's human space flight program, recommended in its final report of 23 October 2009 the extension of the ISS programme to at least 2020.[126] In particular, Leroy Chiao, a former space station commander and space shuttle astronaut who sat on the advisory panel, stated in a CNN interview: "You've got all of these different countries working together on this common project in space. And if we go ahead and stop [...] it's going to break up that framework. The different countries around the world will lose confidence in the US as a leader in space exploration." NASA officials received confirmation from the Obama administration on the future direction of the ISS in particular and the human spaceflight programme in general on 1 February 2010, with a budget proposing an extension to the ISS programme until at least 2020,[9] [127] with talks between ISS partners suggesting that the station could conceivably remain operational until 2028.[128]

NASA has the responsibility to deorbit the ISS. Although *Zvezda* has a propulsion system used for station-keeping, it is not powerful enough for a controlled deorbit. Options for controlled deorbit of the ISS include the use of a modified European Automated Transfer Vehicle or a specially constructed deorbit vehicle.[129] [130]

According to a 2009 report, RKK Energia is considering methods to remove from the station some modules of the Russian Orbital Segment when the end of mission is reached and use them as a basis for a new station, known as the Orbital Piloted Assembly and Experiment Complex. The modules under consideration for removal from the current ISS include the Multipurpose Laboratory Module (MLM), currently scheduled to be launched at the end of 2011, with other Russian modules which are currently planned to be attached to the MLM until 2015, although still currently unfunded. Neither the MLM nor any additional modules attached to it would have reached the end of their useful lives in 2016 or 2020. The report presents a statement from an unnamed Russian engineer who believes that, based on the experience from *Mir*, a thirty-year life should be possible, except for micrometeorite damage, because the Russian modules have been built with on-orbit refurbishment in mind.[131]

Life on board

Crew schedule

The time zone used on board the ISS is Coordinated Universal Time (UTC). The windows are covered at night hours to give the impression of darkness because the station experiences 16 sunrises and sunsets a day. During visiting space shuttle missions, the ISS crew will mostly follow the shuttle's Mission Elapsed Time (MET), which is a flexible time zone based on the launch time of the shuttle mission.[132] [133] Because the sleeping periods between the UTC time zone and the MET usually differ, the ISS crew often has to adjust its sleeping pattern before the space shuttle arrives and after it leaves to shift from one time zone to the other in a practice known as sleep shifting.[134]

A typical day for the crew begins with a wake-up at 06:00, followed by post-sleep activities and a morning inspection of the station. The crew then eats breakfast and takes part in a daily planning conference with Mission Control before starting work at around 08:10. The first scheduled exercise of the day follows, after which the crew continues work until 13:05. Following a one-hour lunch break, the afternoon consists of more exercise and work before the crew carries out its pre-sleep activities beginning at 19:30, including dinner and a crew conference. The scheduled sleep period begins at 21:30. In general, the crew works 10 hours per day on a weekday, and 5 hours on Saturdays, with the rest of the time their own for relaxation, games or work catch-up.[135]

Sleeping in space

The station provides crew quarters for each member of permanent Expedition crews, with two 'sleep stations' in the Russian Orbital Segment and four more, due to be installed in *Tranquillity*, currently spread around the US Orbital Segment. The American quarters are private, approximately person-sized soundproof booths. A crewmember can sleep in them in a tethered sleeping bag, listen to music, use a laptop, and store personal effects in a large drawer or in nets attached to the module's walls. The module also provides a reading lamp, a shelf and a desktop.[136] [137] [138] Visiting crews have no allocated sleep module, and attach a sleeping bag to an available space on a wall—it is possible to sleep floating freely through the station, but this is generally avoided because of the possibility of bumping into sensitive equipment.[139] It is important that crew accommodations are well ventilated. Otherwise, astronauts can wake up oxygen deprived and gasping for air, because a bubble of their own exhaled carbon dioxide has formed around their heads.[138]

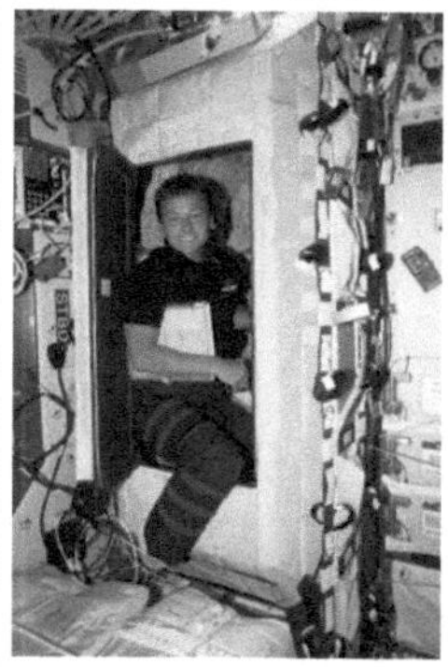

Astronaut Peggy Whitson in the doorway of a sleeping rack in the *Destiny* laboratory

Hygiene

The ISS does not feature a shower, although it was planned as part of the now cancelled Habitation Module. Instead, crewmembers wash using a water jet and wet wipes, with soap dispensed from a toothpaste tube-like container. Crews are also provided with rinseless shampoo and edible toothpaste to save water.[139]

There are two space toilets on the ISS, both of Russian design, located in *Zvezda* and *Destiny*.[136] These Waste and Hygiene Compartments use a fan-driven suction system similar to the Space Shuttle Waste Collection System. Astronauts first fasten themselves to the toilet seat, which is equipped with spring-loaded restraining bars to ensure a good seal.[138] A lever operates a powerful fan and a suction hole slides open: the air stream carries the waste away. Solid waste is collected in individual bags which are stored in an aluminium container. Full containers are transferred to Progress spacecraft for disposal.[136] [140] Liquid waste is evacuated by a hose connected to the front of the toilet, with anatomically correct "urine funnel adapters" attached to the tube so both men and women can use the same toilet. Waste is collected and transferred to the Water Recovery System, where it is recycled back into drinking water.[137]

Food and drink

The crews of STS-127 and Expedition 20 enjoy a meal inside *Unity*.

Most of the food eaten by station crews is frozen, refrigerated or canned. Menus are prepared by the astronauts, with the help of a dietitian, before the astronauts' flight to the station.[137] As the sense of taste is reduced in orbit because of fluid shifting to the head, spicy food is a favourite of many crews.[138] Each crewmember has individual food packages and cooks them using the onboard galley, which features two food warmers, a refrigerator, and a water dispenser that provides both heated and unheated water.[136] Drinks are provided in dehydrated powder form, and are mixed with water before consumption.[136] [137] Drinks and soups are sipped from plastic bags with straws, while solid food is eaten with a knife and fork, which are attached to a tray with magnets to prevent them floating away. Any food which does float away, including crumbs, must be collected to prevent it from clogging up the station's air filters and other equipment.[137]

Exercise

The most significant adverse effects of long-term weightlessness are muscle atrophy and deterioration of the skeleton, or spaceflight osteopenia. Other significant effects include fluid redistribution, a slowing of the cardiovascular system, decreased production of red blood cells, balance disorders, and a weakening of the immune system. Lesser symptoms include loss of body mass, nasal congestion, sleep disturbance, excess flatulence, and puffiness of the face. These effects begin to reverse quickly upon return to the Earth.[36]

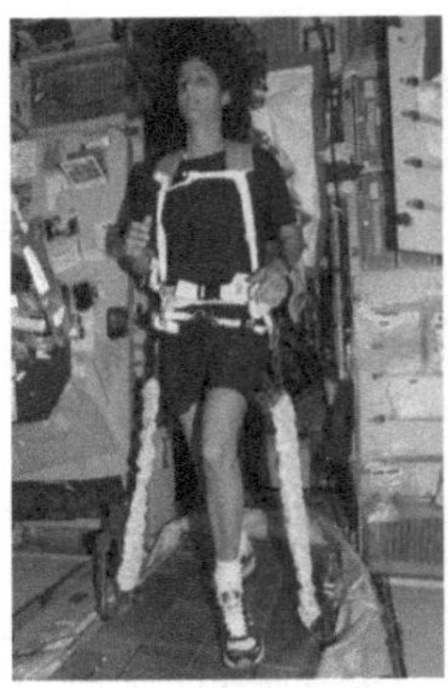

Astronaut Sunita "Suni" Williams is attached to the TVIS treadmill with bungee cords aboard the International Space Station

To prevent some of these adverse physiological effects, the station is equipped with two treadmills (including the COLBERT), the aRED (advanced Resistive Exercise Device) which enables various weightlifting exercises, and a stationary bicycle; each astronaut spends at least two hours per day exercising on the equipment.[136] [138] Astronauts use bungee cords to strap themselves to the treadmill.[141] Researchers believe that exercise is a good countermeasure for the bone and muscle density loss that occurs when humans live for a long time without gravity.[142]

Station operations

Expeditions

Each permanent station crew is given a sequential expedition number. Expeditions have an average duration of half a year, and they commence following the official handover of the station from one Expedition commander to another. Expeditions 1 through 6 consisted of three person crews, but the *Columbia* accident led to a reduction to two crew members for Expeditions 7 to 12. Expedition 13 saw the restoration of the station crew to three, and the station has been permanently staffed as such since. While only three crew members are permanently on the station, several expeditions, such as Expedition 16, have consisted of up to six astronauts or cosmonauts, who are flown to and from the station on separate flights.[143] [144]

On 27 May 2009, Expedition 20 began. Expedition 20 was the first ISS crew of six. Before the expansion of the living volume and capabilities from STS-115 the station could only host a crew of three. Expedition 20's crew was lifted to the station in two separate Soyuz-TMA flights launched at two different times (each Soyuz-TMA can hold only three people): Soyuz TMA-14 on 26 March 2009 and Soyuz TMA-15 on 27 May 2009. However, the station would not be permanently occupied by six crew members all year. For example, when the Expedition 20 crew (Roman Romanenko, Frank De Winne and Bob Thirsk) returned to Earth in November 2009, for a period of about two weeks only two crew members (Jeff Williams and Max Surayev) were aboard. This increased to five in early December, when Oleg Kotov, Timothy Creamer and Soichi Noguchi arrived on Soyuz TMA-17. It decreased to three when Williams and Surayev departed in March 2010, and finally returned to six in April 2010 with the arrival of Soyuz TMA-18, carrying Aleksandr Skvortsov, Mikhail Korniyenko and Tracy Caldwell Dyson.[143] [144]

The International Space Station is the most-visited spacecraft in the history of space flight. As of 24 November 2009, it had had 266 visitors (185 different people).[11] *Mir* had 137 visitors (104 different people).[18]

Visiting spacecraft

The Space Shuttle *Endeavour* approaching the ISS during STS-118

Spacecraft from four different space agencies visit the ISS, serving a variety of purposes. The Automated Transfer Vehicle from the European Space Agency, the Russian Roskosmos Progress spacecraft and the H-II Transfer Vehicle from the Japan Aerospace Exploration Agency have provided resupply services to the station. In addition, Russia supplies a Soyuz spacecraft used for crew rotation and emergency evacuation, which is replaced every six months. Finally, the US services the ISS through its Space Shuttle programme, providing resupply missions, assembly and logistics flights, and crew rotation. As of 27 November 2009, there have been 20 Soyuz, 35 Progress, 1 ATV, 1 HTV and 31 Space Shuttle flights to the station.[2] Expeditions require, on average, 2,722 kg of supplies, and as of 27 November 2009, crews had consumed a total of around 19,000 meals.[2] Soyuz crew rotation flights and Progress resupply flights visit the station on average two and three times respectively each year,[145] with the ATV and HTV planned to visit annually from 2010 onwards.

Following the retirement of the Space Shuttle, a number of other spacecraft are expected to fly to the station. Two, the Orbital Sciences Cygnus and SpaceX Dragon, will fly under NASA's Commercial Orbital Transportation Services and Commercial Resupply Services contracts, delivering cargo to the station until at least 2015.[146] [147] In addition, the Orion spacecraft, developed as a Space Shuttle replacement as part of NASA's Constellation Programme, was retasked by President Barack Obama on 15 April 2010 to provide lifeboat services to the station.[148] The spacecraft had until that point been entirely cancelled under the US 2011 fiscal year budget.[149]

As of 17 June 2010, there were three spacecraft docked with the ISS:[145]

Spacecraft	Mission	Docking port	Date docked (UTC)	Notes
Soyuz TMA-18	Expedition 23/Expedition 24	*Poisk*	5 April 2010 05:24	[17] [150]
Progress M-05M	ISS Progress 37	*Pirs*	1 May 2010 18:30	[151]
Soyuz TMA-19	Expedition 24/Expedition 25	*Zvezda* aft	17 June 2010 21:21	

Mission control centres

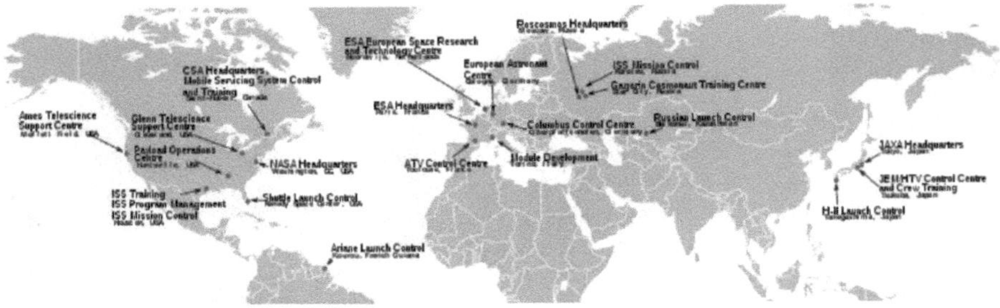

Space centres involved with the ISS programme

The components of the ISS are operated and monitored by their respective space agencies at control centres across the globe, including:

- NASA's Mission Control Center at Lyndon B. Johnson Space Center in Houston, Texas, serves as the primary control facility for the US segment of the ISS and also controls the Space Shuttle missions that visit the station.[24]
- NASA's Payload Operations and Integration Center at Marshall Space Flight Center in Huntsville, Alabama, serves as the centre that coordinates all payload operations in the US Segment.[24]
- Roskosmos's Mission Control Center at Korolyov, Moscow, controls the Russian Orbital Segment of the ISS, in addition to individual Soyuz and Progress missions.[24]
- ESA's Columbus Control Centre at the German Aerospace Centre (DLR) in Oberpfaffenhofen, Germany, controls the European *Columbus* research laboratory.[24]
- ESA's ATV Control Centre, at the Toulouse Space Centre (CST) in Toulouse, France, controls flights of the unmanned European Automated Transfer Vehicle.[24]
- JAXA's JEM Control Centre and HTV Control Centre at Tsukuba Space Centre (TKSC) in Tsukuba, Japan, are responsible for operating the Japanese Experiment Module complex and all flights of the unmanned Japanese H-II Transfer Vehicle respectively.[24]
- CSA's MSS Control at Saint-Hubert, Quebec, Canada, controls and monitors the Mobile Servicing System, or Canadarm2.[24]

Safety aspects

Orbital debris

At the low altitudes at which the ISS orbits there is a variety of space debris, consisting of everything from entire spent rocket stages and defunct satellites, to explosion fragments, paint flakes, slag from solid rocket motors, coolant released by RORSAT nuclear powered satellites, small needles, and many other objects.[152] These objects, in addition to natural micrometeoroids,[153] pose a threat to the station as they have the ability to puncture the pressurised modules and cause damage to other parts of the station.[154] [155] Micrometeoroids also pose a risk to spacewalking astronauts, as such objects could puncture their spacesuits, causing them to depressurise.[156]

The entry hole in Space Shuttle *Endeavour*'s radiator panel caused by space debris during STS-118

Space debris objects are tracked remotely from the ground, and the station crew can be notified of many objects with sufficient size to cause damage on impact. This allows for a Debris Avoidance Manoeuvre (DAM) to be conducted, which uses thrusters on the Russian Orbital Segment to alter the station's orbital altitude, avoiding the debris. DAMs are not uncommon, taking place if computational models show the debris will approach within a certain threat distance.[154] Eight DAMs had been performed prior to March 2009,[157] the first seven between October 1999 and May 2003.[158] Usually the orbit is raised by one or two kilometres by means of an increase in orbital velocity of the order of 1 m/s. Unusually there was a lowering of 1.7 km [158] on 27 August 2008, the first such lowering for 8 years.[159] [160] There were two DAMs in 2009, on 22 March and 17 July.[161] If a threat from orbital debris is identified too late for a DAM to be safely conducted, the station crew close all the hatches aboard the station and retreat into their Soyuz spacecraft, so that they would be able to evacuate in the event it was damaged by the debris. This partial station evacuation has occurred twice, on 6 April 2003 and 13 March 2009.[154]

Radiation

Without the protection of the Earth's atmosphere, astronauts are exposed to higher levels of radiation from a steady flux of cosmic rays. The station's crews are exposed to about 1 millisievert of radiation each day, which is about the same as someone would get from natural sources on Earth in a year.[162] This results in a higher risk of astronauts' developing cancer. High levels of radiation can cause damage to the chromosomes of lymphocytes. These cells are central to the immune system and so any damage to them could contribute to the lowered immunity experienced by astronauts. Over time lowered immunity results in the spread of infection between crew members, especially in such confined areas. Radiation has also been linked to a higher incidence of cataracts in astronauts. Protective shielding and protective drugs may lower the risks to an acceptable level, but data is scarce and longer-term exposure will result in greater risks.[36]

Despite efforts to improve radiation shielding on the ISS compared to previous stations such as *Mir*, radiation levels within the station have not been vastly reduced, and it is thought that further technological advancement will be required to make long-duration human spaceflight further into the Solar System a possibility.[162]

It should be noted, however, that the radiation levels experienced on ISS are not excessively greater than those experienced by airline passengers. The Earth's electromagnetic field provides almost the same level of protection against solar and other radiation in Low Earth Orbit as in the stratosphere. Airline passengers, however, only experience this level of radiation for no more than 15 hours for the longest transcontinental flights (London-Sydney or Chicago-Delhi) and often far less time. For instance, a 12 hour flight from Boston to Beijing, an airline passenger would experience 0.1 millisievert of radiation, or a rate of 0.2 millisieverts per day, only 1/5th the rate experienced by an astronaut in LEO.[163]

External links

Official International Space Station webpages of the participating space agencies

- NASA [164]
- RSC Energia [165]
- Roskosmos [166] (in Russian)
- Canadian Space Agency [167]
- European Space Agency [168]
- Japanese Space Agency [169]
- Italian Space Agency [170]
- Brazilian Space Agency [171] (in Portuguese, archived version)

Interactive and multimedia

- NASA's ISS interactive reference guide [172]
- NASA's ISS image gallery search page [173]
- Current position of the ISS [174]
- ISS WebCam [175]

Experiments and science

- ESA - Columbus [176]
- JAXA - Space Environment Utilization and Space Experiment [177]
- NASA - Station Science [178]
- RSC Energia - Science Research on ISS Russian Segment [179]

References

[1] http://nssdc.gsfc.nasa.gov/nmc/masterCatalog.do?sc=1998-067A

[2] NASA (27 November 2009). "The ISS to Date" (http://spaceflight.nasa.gov/station/isstodate.html). NASA. . Retrieved 28 November 2009.

[3] NASA (29 May 2009). "On-Orbit Elements" (http://www.nasa.gov/externalflash/ISSRG/pdfs/on_orbit.pdf) (PDF). NASA. . Retrieved 28 November 2009.

[4] "ISS Height Profile" (http://www.heavens-above.com/issheight.aspx). Heavens-Above.com. . Retrieved 15 October 2007.

[5] Chris Peat (27 November 2009). "ISS—Orbit Data" (http://www.heavens-above.com/orbit.aspx?satid=25544). Heavens-Above.com. . Retrieved 28 November 2009.

[6] Steven Siceloff (1 February 2001). "NASA Yields to Use of Alpha Name for Station" (http://www.space.com/missionlaunches/missions/fl_alpha_010201.html). *Florida Today*. . Retrieved 18 January 2009.

[7] "Human Space Flight (HSF)—Realtime Data" (http://spaceflight.nasa.gov/realdata/sightings/index.html). NASA. 3 June 2008. . Retrieved 19 December 2009.

[8] Rand Simberg (29 July 2008). "The Uncertain Future of the International Space Station: Analysis" (http://www.popularmechanics.com/science/air_space/4275571.html). Popular Mechanics. . Retrieved 6 March 2009.

[9] NASA (1 February 2010). "Statement by Charlie Bolden, NASA Budget Press Conference" (http://www.nasa.gov/pdf/420994main_2011_Budget_Administrator_Remarks.pdf). Press release. . Retrieved 1 February 2010.

[10] NASA (2 July 2008). "International Space Station Sighting Opportunities" (http://spaceflight.nasa.gov/realdata/sightings/index.html). NASA. . Retrieved 28 January 2009.

[11] "Nations Around the World Mark 10th Anniversary of International Space Station" (http://www.nasa.gov/mission_pages/station/main/10th_anniversary.html). NASA. 17 November 2008. . Retrieved 6 March 2009.

[12] "International Space Station Overview" (http://www.shuttlepresskit.com/ISS_OVR/index.htm). ShuttlePressKit.com. 3 June 1999. . Retrieved 17 February 2009.

[13] "Fields of Research" (http://web.archive.org/web/20080123150641/http://pdlprod3.hosc.msfc.nasa.gov/A-fieldsresearch/index.html). NASA. 26 June 2007. Archived from the original (http://pdlprod3.hosc.msfc.nasa.gov/A-fieldsresearch/index.html) on 25 March 2008. .

[14] "Getting on Board" (http://web.archive.org/web/20071208091537/http://pdlprod3.hosc.msfc.nasa.gov/B-gettingonboard/index.html). NASA. 26 June 2007. Archived from the original (http://pdlprod3.hosc.msfc.nasa.gov/B-gettingonboard/index.html) on 8 December 2007. .

[15] "ISS Research Program" (http://spaceflightsystems.grc.nasa.gov/Advanced/ISSResearch/). NASA. . Retrieved 27 February 2009.

[16] "We've Only Just Begun" (http://www.nasa.gov/mission_pages/station/main/5_year_anniversary.html). NASA. 26 June 2008. . Retrieved 6 March 2009.

[17] "Expedition 23" (http://www.nasa.gov/mission_pages/station/expeditions/expedition23/index.html). NASA. . Retrieved 1 April 2010.

[18] David Harland (30 November 2004). *The Story of Space Station Mir* (http://www.amazon.co.uk/exec/obidos/ASIN/038723011 4/ ref=ord_cart_shr/202-3649698-1866219?_encoding=UTF8&m=A3P5ROKL5A1OLE). New York: Springer-Verlag New York Inc. ISBN 978-0-387-23011-5. .

[19] John E. Catchpole (17 June 2008). *The International Space Station: Building for the Future*. Springer-Praxis. ISBN 978-0387781440.

[20] Kim Dismukes (4 April 2004). "Shuttle–Mir History/Background/How "Phase 1" Started" (http://spaceflight.nasa.gov/history/ shuttle-mir/history/h-b-start.htm). NASA. . Retrieved 12 April 2007.

[21] "Spread Your Wings, It's Time to Fly" (http://www.nasa.gov/mission_pages/station/behindscenes/truss_segment.html). NASA. 26 July 2006. . Retrieved 21 September 2006.

[22] NASA (15 December 2008). "Current ISS Tracking data" (http://spaceflight.nasa.gov/realdata/tracking/index.html). NASA. . Retrieved 28 January 2009.

[23] "Human Spaceflight and Exploration—European Participating States" (http://www.esa.int/esaHS/partstates.html). European Space Agency (ESA). 2009. . Retrieved 17 January 2009.

[24] Gary Kitmacher (2006). *Reference Guide to the International Space Station*. Canada: Apogee Books. pp. 71–80. ISBN 978-1-894959-34-6.

[25] "ISS Intergovernmental Agreement" (http://www.spaceflight.esa.int/users/index.cfm?act=default.page&level=11&page=1980). European Space Agency (ESA). 19 April 2009. . Retrieved 19 April 2009.

[26] "Memorandum of Understanding Between the National Aeronautics and Space Administration of the United States of America and the Russian Space Agency Concerning Cooperation on the Civil International Space Station" (http://www.nasa.gov/mission_pages/station/ structure/elements/nasa_rsa.html). NASA. 29 January 1998. . Retrieved 19 April 2009.

[27] "How Much Does It Cost?" (http://www.esa.int/esaHS/ESAQHA0VMOC_iss_0.html). European Space Agency (ESA). 9 August 2005. . Retrieved 27 March 2008.

[28] Alan Boyle (25 August 2006). "What's the cost of the space station?" (http://www.msnbc.msn.com/id/14505278/). MSNBC. . Retrieved 30 September 2008.

[29] Jim Wilson (December 2002). "Scientists Believe ISS Is Waste Of Money" (http://www.popularmechanics.com/science/air_space/ 1282806.html). *Popular Mechanics*. Hearst Corporation. . Retrieved 17 December 2009.

[30] James P. Bagian *et al.* (2001). *Readiness Issues Related to Research in the Biological and Physical Sciences on the International Space Station* (http://www.nap.edu/catalog.php?record_id=10196). United States National Academy of Sciences. .

[31] James Oberg (2005). "International Space Station" (http://www.nasa.gov/worldbook/intspacestation_worldbook.html). *World Book Online Reference Center*. World Book, Inc. . Retrieved 14 June 2008.

[32] "The International Space Station: life in space" (http://www.scienceinschool.org/2008/issue10/iss). Science in School. 10 December 2008. . Retrieved 17 February 2009.

[33] Chris Bergin (22 August 2009). "ISS: Still in assembly, producing science research accomplishments" (http://www.nasaspaceflight.com/ 2009/08/iss-assembly-producing-science-research-accomplishments/). NASASpaceflight.com. . Retrieved 27 September 2009.

[34] Gro Mjeldheim Sandal and Dietrich Manzey (December 2009). "Cross-cultural issues in space operations: A survey study among ground personnel of the European Space Agency". *Acta Astronautica* **65**: 1520–1529. doi:10.1016/j.actaastro.2009.03.074.

[35] "NASA Authorization Act 2005" (http://frwebgate.access.gpo.gov/cgi-bin/getdoc.cgi?dbname=109_cong_public_laws& docid=f:publ155.109.pdf). United States Government Printing Office. 30 December 2005. . Retrieved 6 March 2009.

[36] Jay Buckey (23 February 2006). *Space Physiology* (http://www.amazon.co.uk/Space-Physiology-Jay-C-Buckey/dp/0195137256/ ref=sr_1_1?ie=UTF8&s=books&qisbn=1-227-46880-9&sr=8-1). Oxford University Press USA. ISBN 978-0-19-513725-5. .

[37] List Grossman (24 July 2009). "Ion engine could one day power 39-day trips to Mars" (http://www.newscientist.com/article/ dn17476-ion-engine-could-one-day-power-39day-trips-to-mars.html?full=true). New Scientist. . Retrieved 8 January 2010.

[38] Brooke Boen (1 May 2009). "Advanced Diagnostic Ultrasound in Microgravity (ADUM)" (http://www.nasa.gov/mission_pages/station/ science/experiments/ADUM.html). NASA. . Retrieved 1 October 2009.

[39] Sishir Rao *et al.* (2008). "A Pilot Study of Comprehensive Ultrasound Education at the Wayne State University School of Medicine" (http:// www.jultrasoundmed.org/cgi/content/abstract/27/5/745). *Journal of Ultrasound in Medicine* **27** (5): 745–749. PMID 18424650. .

[40] Michael Fincke *et al.* (2004). "Evaluation of Shoulder Integrity in Space: First Report of Musculoskeletal US on the International Space Station" (http://radiology.rsna.org/content/234/2/319.abstract). *Radiology* **234** (234): 319–322. doi:10.1148/radiol.2342041680. PMID 15533948. .

[41] "Materials Science 101" (http://science.nasa.gov/newhome/headlines/msad15sep99_1.htm). Science@NASA. 15 September 1999. . Retrieved 18 June 2009.

[42] Donna Heivilin (21 June 1994). "*Space Station: Impact of the Expanded Russian Role on Funding and Research*" (http://archive.gao.gov/ t2pbat3/151975.pdf) (PDF). Government Accountability Office. . Retrieved 3 November 2006.

[43] NASA (2008). "Consolidated Launch Manifest" (http://www.nasa.gov/mission_pages/station/structure/iss_manifest.html). NASA. . Retrieved 8 July 2008.

[44] "HSF: ISS assembly sequence and on-orbit configuration" (http://esamultimedia.esa.int/multimedia/esa_iss_assembly_sequence/ index_pop.html). European Space Agency (ESA). . Retrieved 6 March 2009.

[45] Chris Bergin (26 July 2005). "Discovery launches—The Shuttle is back" (http://www.nasaspaceflight.com/2005/07/discovery-launches-the-shuttle-is-back/). NASASpaceflight.com. . Retrieved 6 March 2009.
[46] "Zarya Module" (http://www.nasa.gov/mission_pages/station/structure/elements/fgb.html). NASA. 14 October 2008. . Retrieved 7 December 2009.
[47] "Unity Connecting Module: Cornerstone for a Home in Orbit" (http://spaceflight.nasa.gov/spacenews/factsheets/pdfs/unity.pdf). NASA. January 1999. . Retrieved 11 March 2009.
[48] "Zvezda Service Module" (http://www.nasa.gov/mission_pages/station/structure/elements/sm.html). NASA. 11 March 2009. . Retrieved 11 March 2009.
[49] "NASA—US Destiny Laboratory" (http://www.nasa.gov/mission_pages/station/structure/elements/destiny.html). NASA. 26 March 2007. . Retrieved 26 June 2007.
[50] "Space Station Extravehicular Activity" (http://spaceflight.nasa.gov/station/eva/outside.html). NASA. 4 April 2004. . Retrieved 11 March 2009.
[51] "Pirs Docking Compartment" (http://www.nasa.gov/mission_pages/station/structure/elements/pirs.html). NASA. 10 May 2006. . Retrieved 28 March 2009.
[52] "Harmony Node 2" (http://www.nasa.gov/mission_pages/station/structure/elements/node2.html). NASA. 26 September 2007. . Retrieved 28 March 2009.
[53] Chris Bergin (10 January 2008). "PRCB plan STS-122 for NET Feb 7—three launches in 10–11 weeks" (http://www.nasaspaceflight.com/2008/01/prcb-plan-sts-122-for-net-feb-7-three-launches-in-10-11-weeks/). NASASpaceflight.com. . Retrieved 12 January 2008.
[54] "Columbus laboratory" (http://www.esa.int/esaHS/ESAAYI0VMOC_iss_0.html). European Space Agency (ESA). 10 January 2009. . Retrieved 6 March 2009.
[55] "NASA—Kibo Japanese Experiment Module" (http://www.nasa.gov/mission_pages/station/structure/elements/jem.html). NASA. 23 November 2007. . Retrieved 28 March 2009.
[56] "About Kibo" (http://kibo.jaxa.jp/en/about/). Japan Aerospace Exploration Agency (JAXA). 25 September 2008. . Retrieved 6 March 2009.
[57] Anatoly Zak. "Docking Compartment-1 and 2" (http://www.russianspaceweb.com/iss_dc.html). RussianSpaceWeb.com. . Retrieved 26 March 2009.
[58] Chris Bergin (10 November 2009). "Russian module launches via Soyuz for Thursday ISS docking" (http://www.nasaspaceflight.com/2009/11/live-russian-module-launch-towards-iss-on-soyuz/). NASASpaceflight.com. . Retrieved 10 November 2009.
[59] Robert Z. Pearlman (15 April 2009). "NASA Names Space Module After Moon Base, Not Stephen Colbert" (http://www.space.com/news/090414-colbert-space-station-node.html). Space.com. . Retrieved 15 April 2009.
[60] "Node 3: Connecting Module" (http://www.esa.int/esaHS/ESAFQL0VMOC_iss_0.html). European Space Agency (ESA). 23 February 2009. . Retrieved 28 March 2009.
[61] "Cupola" (http://www.esa.int/esaHS/ESA65K0VMOC_iss_0.html). European Space Agency (ESA). 16 January 2009. . Retrieved 28 March 2009.
[62] Chris Gebhardt (5 August 2009). "STS-133 refined to a five crew, one EVA mission—will leave MPLM on ISS" (http://www.nasaspaceflight.com/2009/08/sts-133-five-crew-one-eva-mission-leave-mpm-on-iss). NASASpaceflight.com. .
[63] Amos, Jonathan (29 August 2009). "Europe looks to buy Soyuz craft" (http://news.bbc.co.uk/2/hi/science/nature/8226309.stm). BBC News. .
[64] "Shuttle Q&A Part 5" (http://forum.nasaspaceflight.com/index.php?topic=17437.msg483604#msg483604). NASASpaceflight.com. 27 September 2009. . Retrieved 12 October 2009.
[65] "FGB-based Multipurpose Lab Module (MLM)" (http://web.archive.org/web/20070927002737/http://www.khrunichev.ru/khrunichev_eng/live/full_mks.asp?id=13190). Khrunichev State Research and Production Space Centre. Archived from the original (http://www.khrunichev.ru/khrunichev_eng/live/full_mks.asp?id=13190) on 27 September 2007. . Retrieved 31 October 2008.
[66] "Where is the Centrifuge Accommodation Module (CAM)?" (http://forum.nasaspaceflight.com/forums/thread-view.asp?tid=12560&mid=269666). NASASpaceflight.com. . Retrieved 12 October 2009.
[67] Tariq Malik (14 February 2006). "NASA Recycles Former ISS Module for Life Support Research" (http://www.space.com/missionlaunches/060214_iss_module.html). Space.com. . Retrieved 11 March 2009.
[68] E. D. Graf (February 2000). "The X-38 and Crew Return Vehicle Programmes" (http://esapub.esrin.esa.it/bulletin/bullet101/graf.pdf) (PDF). *ESA Bulletin 101*. European Space Agency. . Retrieved 4 October 2009.
[69] "ICM Interim Control Module" (http://web.archive.org/web/20070208164211/http://code8200.nrl.navy.mil/icm.html). U.S. Naval Center for Space Technology. Archived from the original (http://code8200.nrl.navy.mil/icm.html) on 8 February 2007. .
[70] Anatoly Zak. "Russian segment of the ISS" (http://www.russianspaceweb.com/iss_russia.html). russianspaceweb.com. . Retrieved 3 October 2009.
[71] "Russian Research Modules" (http://www.boeing.com/defense-space/space/spacestation/components/russian_laboratory.html). Boeing. . Retrieved 21 June 2009.
[72] "The Alpha Magnetic Spectrometer Experiment" (http://ams.cern.ch/). CERN. 21 January 2009. . Retrieved 6 March 2009.
[73] "International Space Station" (http://www.asc-csa.gc.ca/eng/iss/default.asp). Canadian Space Agency. 9 March 2006. . Retrieved 4 October 2009.

[74] Chris Bergin (27 January 2009). "NASA approve funding to leave OBSS permanently on the ISS" (http://www.nasaspaceflight.com/2009/01/nasa-approve-leave-obss-permanently-on-iss/). NASASpaceflight.com. . Retrieved 5 October 2009.
[75] "ERA: European Robotic Arm" (http://www.esa.int/esaHS/ESAQEI0VMOC_iss_0.html). ESA. . Retrieved 4 October 2009.
[76] "Remote Manipulator System:About Kibo" (http://kibo.jaxa.jp/en/about/kibo/rms/). JAXA. 29 August 2008. . Retrieved 4 October 2009.
[77] "International Space Station Status Report #02-03" (http://www.nasa.gov/centers/johnson/news/station/2002/iss02-03.txt). NASA. 14 January 2002. . Retrieved 4 October 2009.
[78] "EXPRESS Racks 1 and 2 fact sheet" (http://www.nasa.gov/centers/marshall/news/background/facts/expressrack.html). NASA. 12 April 2008. . Retrieved 4 October 2009.
[79] "Exposed Facility:About Kibo" (http://kibo.jaxa.jp/en/about/kibo/jef/). JAXA. 29 August 2008. . Retrieved 9 October 2009.
[80] "NASA—European Technology Exposure Facility (EuTEF)" (http://www.nasa.gov/mission_pages/station/science/experiments/EuTEF.html). NASA. 6 October 2008. . Retrieved 28 February 2009.
[81] "ESA—Columbus—European Technology Exposure Facility (EuTEF)" (http://www.esa.int/esaMI/Columbus/SEM7ZTEMKBF_0.html). ESA. 13 January 2009. . Retrieved 28 February 2009.
[82] "Atomic Clock Ensemble in Space (ACES)" (http://www.spaceflight.esa.int/projects/index.cfm?act=default.page&level=12&page=829). ESA. . Retrieved 9 October 2009.
[83] "Boeing: Integrated Defense Systems—NASA Systems—International Space Station—Solar Power" (http://www.boeing.com/defense-space/space/spacestation/systems/solar_arrays.html). Boeing. 2 November 2006. . Retrieved 28 January 2009.
[84] Thomas B. Miller (24 April 2000). "Nickel-Hydrogen Battery Cell Life Test Program Update for the International Space Station" (http://www.grc.nasa.gov/WWW/RT/RT1999/5000/5420miller.html). NASA. . Retrieved 27 November 2009.
[85] G. Landis & C-Y. Lu (1991). "Solar Array Orientation Options for a Space Station in Low Earth Orbit". *Journal of Propulsion and Power* **7** (1): 123–125. doi:10.2514/3.23302.
[86] "ISS Environment" (http://web.archive.org/web/20080213164432/http://pdlprod3.hosc.msfc.nasa.gov/D-aboutiss/D6.html). Johnson Space Center. Archived from the original (http://pdlprod3.hosc.msfc.nasa.gov/D-aboutiss/D6.html) on 13 February 2008. . Retrieved 15 October 2007.
[87] "Press Release 121208" (http://www.adastrarocket.com/AdAstra-NASA_PR12Dec08.pdf) (pdf). AdAstra Rocket Company. 12 December 2008. . Retrieved 7 December 2009.
[88] "Propulsion Systems of the Future" (http://www.nasa.gov/vision/space/travelinginspace/future_propulsion.html). NASA. . Retrieved 29 May 2009.
[89] David Shiga (5 October 2009). "Rocket company tests world's most powerful ion engine" (http://www.newscientist.com/article/dn17918-rocket-company-tests-worlds-most-powerful-ion-engine.html). New Scientist. . Retrieved 7 October 2009.
[90] Carlos Roithmayr (2003). *Dynamics and Control of Attitude, Power, and Momentum for a Spacecraft Using Flywheels and Control Moment Gyroscopes* (http://www.amazon.co.uk/Dynamics-Attitude-Spacecraft-Flywheels-Gyroscopes/dp/B0018V9YG6/ref=sr_1_1?ie=UTF8&s=books&qid=1236365448&sr=8-1). Langley Research Centre: NASA. .
[91] "International Space Station Status Report #05-7" (http://spaceflight.nasa.gov/spacenews/reports/issreports/2005/iss05-7.html). NASA. 11 February 2005. . Retrieved 23 November 2008.
[92] Chris Bergin (14 June 2007). "Atlantis ready to support ISS troubleshooting" (http://www.nasaspaceflight.com/2007/06/atlantis-ready-to-support-iss-troubleshooting/). NASASPaceflight.com. . Retrieved 6 March 2009.
[93] "Communications and Tracking" (http://web.archive.org/web/20080611115319/http://www.boeing.com/defense-space/space/spacestation/systems/communications_tracking.html). Boeing. Archived from the original (http://www.boeing.com/defense-space/space/spacestation/systems/communications_tracking.html) on June 11, 2008. . Retrieved 30 November 2009.
[94] Mathews, Melissa; James Hartsfield (25 March 2005). "International Space Station Status Report: SS05-015" (http://www.nasa.gov/home/hqnews/2005/mar/HQ_ss05015_ISS_status_report.html). *NASA News*. NASA. . Retrieved 11 January 2010.
[95] Harvey, Brian (2007). *The rebirth of the Russian space program: 50 years after Sputnik, new frontiers*. Springer Praxis Books. pp. 263. ISBN 0387713549.
[96] Anatoly Zak (4 January 2010). "Space exploration in 2011" (http://www.russianspaceweb.com/2011.html). RussianSpaceWeb. . Retrieved 12 January 2010.
[97] "Memorandum of Understanding Between the National Aeronautics and Space Administration of the United States of America and the Government of Japan Concerning Cooperation on the Civil International Space Station" (http://www.nasa.gov/mission_pages/station/structure/elements/nasa_japan.html). NASA. 24 February 1998. . Retrieved 19 April 2009.
[98] "Operations Local Area Network (OPS LAN) Interface Control Document" (http://www.spaceref.com/iss/computer/iss.ops.lan.icd.pdf) (PDF). NASA. February 2000. . Retrieved 30 November 2009.
[99] "ISS/ATV communication system flight on Soyuz" (http://www.spaceref.com/news/viewpr.html?pid=16247). EADS Astrium. 28 February 2005. . Retrieved 30 November 2009.
[100] Chris Bergin (10 November 2009). "STS-129 ready to support Dragon communication demo with ISS" (http://www.nasaspaceflight.com/2009/11/sts-129-support-dragon-communication-demo-iss/). NASASpaceflight.com. . Retrieved 30 November 2009.
[101] "Microgravity" (http://www.bnsc.gov.uk/Discovering Space/Microgravity/8057.aspx) (aspx). British National Space Council. 3 August 2009. . Retrieved 7 September 2009.

[102] "European Users Guide to Low Gravity Platforms" (http://www.spaceflight.esa.int/users/downloads/userguides/physenv.pdf) (PDF). European Space Agency. 6 December 2005. . Retrieved 16 May 2006.

[103] Tariq Malik (15 February 2006). "Air Apparent: New Oxygen Systems for the ISS" (http://www.space.com/businesstechnology/060215_techwed_iss_oxygen.html). Space.com. . Retrieved 21 November 2008.

[104] Patrick L. Barry (13 November 2000). "Breathing Easy on the Space Station" (http://science.nasa.gov/headlines/y2000/ast13nov_1.htm). NASA. . Retrieved 21 November 2008.

[105] Craig Freudenrich (20 November 2000). "How Space Stations Work" (http://science.howstuffworks.com/space-station2.htm). Howstuffworks. . Retrieved 23 November 2008.

[106] "5–8: The Air Up There" (http://web.archive.org/web/20061114010931/http://www.nasaexplores.com/show2_5_8a.php?id=04-032&gl=58). *NASAexplores*. NASA. Archived from the original (http://nasaexplores.com/show2_5_8a.php?id=04-032&gl=58) on 14 November 2006. . Retrieved 31 October 2008.

[107] Clinton Anderson; et. al. (30 January 1968). *Report of the Committee on Aeronautical and Space Sciences, United States Senate—Apollo 204 Accident* (http://klabs.org/richcontent/Reports/Failure_Reports/as-204/senate_956/as204_senate_956.pdf). Washington, DC: US Government Printing Office. pp. 8. .

[108] European Space Agency (7 January 2009). "See the ISS from your home town" (http://www.esa.int/esaHS/ESA0I6KE43D_iss_0.html). European Space Agency. . Retrieved 13 November 2007.

[109] Spaceweather.com (5 June 2009). "ISS visible during the daytime" (http://spaceweather.com/archive.php?view=1&day=05&month=06&year=2009). Spaceweather.com. . Retrieved 5 June 2009.

[110] "NASA Signs International Space Station Agreement With Brazil" (http://www.nasa.gov/centers/johnson/news/releases/1996_1998/h97-233.html). NASA. 14 October 1997. . Retrieved 18 January 2009.

[111] "International Space Station (ISS)" (http://www.asi.it/en/flash_en/living/the_international_space_station_iss). Italian Space Agency. 18 January 2009. . Retrieved 18 January 2009.

[112] "China wants role in space station" (http://web.archive.org/web/20080314104006/http://www.cnn.com/2007/TECH/space/10/16/china.space.ap/index.html). Associated Press. CNN. 16 October 2007. Archived from the original (http://www.cnn.com/2007/TECH/space/10/16/china.space.ap/index.html) on 14 March 2008. . Retrieved 20 March 2008.

[113] James Oberg (26 October 2001). "China takes aim at the space station" (http://www.msnbc.msn.com/id/3077826/). MSNBC. . Retrieved 30 January 2009.

[114] "South Korea, India to begin ISS partnership talks in 2010" (http://www.flightglobal.com/articles/2009/10/14/333406/south-korea-india-to-begin-iss-partnership-talks-in.html). Flight International. October 14, 2009. . Retrieved 2009-10-14.

[115] "Memorandum of Understanding Between the National Aeronautics and Space Administration of the United States of America and the Canadian Space Agency Concerning Cooperation on the Civil International Space Station" (http://www.nasa.gov/mission_pages/station/structure/elements/nasa_csa.html). NASA. 29 January 1998. . Retrieved 19 April 2009.

[116] Robert L. Park. "Space Station: Maybe They Could Use It to Test Missile Defense" (http://bobpark.physics.umd.edu/WN04/wn092404.html). University of Maryland. . Retrieved 23 March 2009.

[117] Bob Park. "Space: International Space Station Unfurls New Solar Panels" (http://bobpark.physics.umd.edu/WN06/wn091506.html). University of Maryland. . Retrieved 15 June 2007.

[118] NASA (2007). "Renal Stone Risk During Spaceflight: Assessment and Countermeasure Validation (Renal Stone)" (http://www.nasa.gov/mission_pages/station/science/experiments/Renal-Stone.html). NASA. . Retrieved 13 November 2007.

[119] NASA (2007). "Sleep-Wake Actigraphy and Light Exposure During Spaceflight-Long (Sleep-Long)" (http://www.nasa.gov/mission_pages/station/science/experiments/Sleep-Long.html). NASA. . Retrieved 13 November 2007.

[120] NASA (2007). "Anomalous Long Term Effects in Astronauts' Central Nervous System (ALTEA)" (http://www.nasa.gov/mission_pages/station/science/experiments/ALTEA.html). NASA. . Retrieved 13 November 2007.

[121] James J. Secosky, George Musser (1996). "Up, Up, and Away" (http://www.astrosociety.org/education/publications/tnl/34/space2.html). Astronomical Society of the Pacific. . Retrieved 10 September 2006.

[122] E. Ginzburg, J.W. Kuhn, J. Schnee & B. Yavitz (1976). "Economic impact of large public programs The NASA experience" (http://ntrs.nasa.gov/index.jsp?method=order&oaiID=19760016995). NASA Technical Reports Server (NTRS). . Retrieved 13 November 2007.

[123] Federation of American Scientists. "NASA Technological Spinoff Fables" (http://www.fas.org/spp/eprint/jp_950525.htm). Federation of American Scientists. . Retrieved 17 September 2006.

[124] Achenbach, Joel (13 July 2009). "As Space Station Nears Completion, It Faces End of Mission" (http://www.washingtonpost.com/wp-dyn/content/article/2009/07/12/AR2009071201977.html). *The Washington Post*. . Retrieved 18 July 2009.

[125] "New NASA boss: Astronauts on Mars in his lifetime" (http://abcnews.go.com/Technology/wireStory?id=8140312). ABC News. 21 July 2009. . Retrieved 7 December 2009.

[126] Gen. Norman Augustine, *et al.* (23 October 2009). "Seeking a Human Spaceflight Program Worthy of a Great Nation" (http://www.nasa.gov/pdf/396093main_HSF_Cmte_FinalReport.pdf). Review of US Human Space Flight Plans Committee/NASA. . Retrieved 6 December 2009.

[127] Rich Phillips (14 October 2009). "How much longer will the space station fly?" (http://www.cnn.com/2009/TECH/space/10/12/space.station.future/index.html). CNN. . Retrieved 14 October 2009.

[128] Stephen Clark (11 March 2010). "Space station partners set 2028 as certification goal" (http://spaceflightnow.com/news/n1003/11station/). Spaceflight Now. . Retrieved 14 March 2010.

[129] Rob Coppinger (7 March 2007). "NASA may buy ESA's ATV to de-orbit ISS at end of life" (http://www.flightglobal.com/articles/2007/07/03/215291/nasa-may-buy-esas-atv-to-de-orbit-iss-at-end-of-life.html). Flightglobal. . Retrieved 27 September 2009.
[130] Thomas Kelly, *et al.* (2000). *Engineering Challenges to the Long-Term Operation of the International Space Station* (http://search.nap.edu/openbook.php?record_id=9794&page=28). National Academies Press. pp. 28–30. ISBN 0-309-06938-6. .
[131] "Russia 'to save its ISS modules'" (http://news.bbc.co.uk/2/hi/science/nature/8064060.stm). BBC News. 22 May 2009. . Retrieved 23 May 2009.
[132] "Mission Elapsed Time explained" (http://web.archive.org/web/20071115223702/http://science.nasa.gov/Realtime/Rocket_Sci/clocks/time-met.html). NASA. 13 September 1995. Archived from the original (http://science.nasa.gov/Realtime/Rocket_Sci/clocks/time-met.html) on November 15, 2007. . Retrieved 9 November 2007.
[133] "Ask the STS-113 crew: Question 14" (http://spaceflight.nasa.gov/feedback/expert/answer/crew/sts-113/index_2.html). NASA. 7 December 2002. . Retrieved 9 November 2007.
[134] "STS-113 Mission and Expedition Crew Question and Answer Board" (http://www.nasa.gov/missions/highlights/webcasts/shuttle/sts113/mission-crew-qa.html). NASA. November 2002. . Retrieved 24 February 2009.
[135] "ISS Crew Timeline" (http://www.nasa.gov/pdf/287386main_110508_tl.pdf) (PDF). NASA. 5 November 2008. . Retrieved 5 November 2008.
[136] Cheryl L. Mansfield (7 November 2008). "Station Prepares for Expanding Crew" (http://www.nasa.gov/mission_pages/station/behindscenes/126_payload.html). NASA. . Retrieved 17 September 2009.
[137] "Living and Working on the International Space Station" (http://www.asc-csa.gc.ca/pdf/educator-liv_wor_iss.pdf). CSA. . Retrieved 28 October 2009.
[138] "Daily life" (http://www.esa.int/esaHS/ESAH1V0VMOC_astronauts_0.html). ESA. 19 July 2004. . Retrieved 28 October 2009.
[139] Tariq Malik (27 July 2009). "Sleeping in Space is Easy, But There's No Shower" (http://www.space.com/missionlaunches/090827-sts127-space-sleeping.html). Space.com. . Retrieved 29 October 2009.
[140] Ed Lu (8 September 2003). ""Greetings Earthling"" (http://spaceflight.nasa.gov/station/crew/exp7/luletters/lu_letter9.html). NASA. . Retrieved 1 November 2009.
[141] "Bungee Cords Keep Astronauts Grounded While Running" (http://www.nasa.gov/mission_pages/station/behindscenes/bungee_running.html). NASA. 16 June 2009. . Retrieved 23 August 2009.
[142] "Do Tread on Me" (http://www.nasa.gov/mission_pages/station/behindscenes/colbert_feature.html). NASA. 19 August 2009. . Retrieved Augist 23, 2009.
[143] "International Space Station Expeditions" (http://www.nasa.gov/mission_pages/station/expeditions/index.html). NASA. 10 April 2009. . Retrieved 13 April 2009.
[144] NASA (2008). "International Space Station" (http://www.nasa.gov/mission_pages/station/main/index.html). NASA. . Retrieved 22 October 2008.
[145] "Live listing of spacecraft operations" (http://www.nasa.gov/mission_pages/station/resupply/index.html). NASA. 1 December 2009. . Retrieved 8 December 2009.
[146] Space Operations Mission Directorate (30 August 2006). "Human Space Flight Transition Plan" (http://www.hq.nasa.gov/osf/tran_plan.pdf). NASA. .
[147] NASA (18 January 2006). "NASA Seeks Proposals for Crew and Cargo Transportation to Orbit" (http://www.spaceref.com/news/viewpr.html?pid=18791). Press release. . Retrieved 21 November 2006.
[148] Barack Obama. "Remarks By the President on Space Exploration in the 21st Centure" (http://en.wikisource.org/wiki/President_Barack_Obama_on_Space_Exploration_in_the_21st_Century) John F. Kennedy Space Centre (15 April 2010). Retrieved on 5 May 2010.
[149] "Fiscal Year 2011 Budget Estimates" (http://www.nasa.gov/pdf/420990main_FY_201_ Budget_Overview_1_Feb_2010.pdf). NASA. . Retrieved 7 March 2010.
[150] Chris Bergin (1 April 2010). "LIVE: Soyuz TMA-18 launches towards an over-stowed ISS" (http://www.nasaspaceflight.com/2010/04/live-soyuz-tma-18-set-to-launch-towards-an-over-stowed-iss/). NASASpaceflight.com. . Retrieved 23 December 2009.
[151] Chris Bergin (1 May 2010). "Progress M-05M replenishes ISS – NTA back to full working order" (http://www.nasaspaceflight.com/2010/05/progress-m-05m-replenishes-iss-nta-full-working-order/). NASASpaceflight.com. . Retrieved 4 May 2010.
[152] Michael Hoffman (3 April 2009). "National Space Symposium 2009:It's getting crowded up there" (http://defensenews.com/blogs/space-symposium/2009/04/03/its-getting-crowded-up-there/#more-155). Defense News. . Retrieved 7 October 2009.
[153] F. L. Whipple (1949). "The Theory of Micrometeoroids" (http://adsabs.harvard.edu/full/1949PA.....57..517W). *Popular Astronomy* **57**: 517. .
[154] Chris Bergin (30 September 2009). "Soyuz TMA-16 launches for journey to ISS—Safe Haven evaluations" (http://www.nasaspaceflight.com/2009/09/soyuz-tma-16-launch-to-iss-safe-haven-evaluations/). NASASpaceflight.com. . Retrieved 7 October 2009.
[155] Henry Nahra (24–29 April 1989). "Effect of Micrometeoroid and Space Debris Impacts on the Space Station Freedom Solar Array Surfaces" (http://ntrs.nasa.gov/archive/nasa/casi.ntrs.nasa.gov/19890016664_1989016664.pdf). NASA. . Retrieved 7 October 2009.
[156] Leonard David (7 January 2002). "Space Junk and ISS: A Threatening Problem" (http://www.space.com/missionlaunches/junk_iss_020107.html). Space.com. . Retrieved 30 November 2008.
[157] Rachel Courtland (16 March 2009). "Space station may move to dodge debris" (http://www.newscientist.com/article/dn16777-space-station-may-move-to-dodge-debris.html). New Scientist. . Retrieved 20 April 2010.

[158] "ISS Maneuvers to Avoid Russian Fragmentation Debris" (http://www.orbitaldebris.jsc.nasa.gov/newsletter/pdfs/ODQNv12i4.pdf). *Orbital Debris Quarterly News* (NASA) **12** (4): 1&2. October 2008. . Retrieved 20 April 2010.
[159] "ATV carries out first debris avoidance manoeuvre for the ISS" (http://www.esa.int/esaMI/ATV/SEM64X0SAKF_0.html). ESA. 28 August 2008. . Retrieved 26 February 2010.
[160] "ISS dodges debris" (http://www.nowpublic.com/tech-biz/iss-dodges-debris). NowPublic. 29 August 2008. . Retrieved 20 April 2010.
[161] "Avoiding satellite collisions in 2009." (http://www.orbitaldebris.jsc.nasa.gov/newsletter/pdfs/ODQNv14i1.pdf). *Orbital Debris Quarterly News* (NASA) **14** (1): 2. January 2010. . Retrieved 20 April 2010.
[162] Eugenie Samuel (23 October 2002). "Space station radiation shields 'disappointing'" (http://www.newscientist.com/article/dn2956-space-station-radiation-shields-disappointing.html). New Scientist. . Retrieved 7 October 2009.
[163] "Galactic Radiation Received in Flight" (http://jag.cami.jccbi.gov./cariprofile.asp). FAA Civil Aeromedical Institute. . Retrieved 20 May 2010.
[164] http://www.nasa.gov/mission_pages/station/main/
[165] http://www.energia.ru/eng/iss/iss.html
[166] http://www.roscosmos.ru/MIOP1file.asp
[167] http://www.asc-csa.gc.ca/eng/iss/default.asp
[168] http://www.esa.int/esaHS/iss.html
[169] http://www.jaxa.jp/projects/iss_human/index_e.html
[170] http://www.asi.it/en/flash_en/living
[171] http://web.archive.org/web/20080801141159/http://www.aeb.gov.br/conteudo.php?ida=28&idc=118
[172] http://www.nasa.gov/externalflash/ISSRG/index.html
[173] http://spaceflight.nasa.gov/gallery/images/station/
[174] http://spaceflight.nasa.gov/realdata/tracking/
[175] http://www.nasa.gov/multimedia/isslivestream.asx
[176] http://www.esa.int/SPECIALS/Columbus/index.html
[177] http://www.jaxa.jp/projects/iss_human/research/index_e.html
[178] http://www.nasa.gov/mission_pages/station/science/index.html
[179] http://www.energia.ru/en/iss/researches/iss-researches.html

Dennis Tito

Dennis Anthony Tito	

Spaceflight Participant	
Born	August 8, 1940 Queens, New York
Other occupation	Entrepreneur
Time in space	7d 22h 04m
Missions	Soyuz TM-32, Soyuz TM-31
Mission insignia	

Dennis Anthony Tito (born August 8, 1940 in Queens, New York) is a United States engineer, multimillionaire, and most widely known as the first space tourist to pay for his own ticket. In mid-2001, he spent nearly eight days in orbit on Soyuz TM-32, the International Space Station, and Soyuz TM-31.

Career

Tito has a Bachelor of Science in Astronautics and Aeronautics from New York University, 1962 and a Master of Science in Engineering Science from Rensselaer Polytechnic Institute satellite campus in Hartford, Connecticut.[1] He is a member of Psi Upsilon and received an honorary doctorate of engineering from Rensselaer Polytechnic Institute on 18 May, 2002 and is a former scientist of the NASA Jet Propulsion Laboratory. In 1972, he founded Wilshire Associates, a leading provider of investment management, consulting and technology services in Santa Monica, California. Tito serves an international clientèle representing assets of $12.5 trillion.[2] Wilshire relies on the field of quantitative analytics, which uses mathematical tools to analyze market risks - a methodology Tito is credited with helping to develop by applying the same techniques he used to determine a spacecraft's path at JPL.[3] Despite a career change from aerospace engineering to investment management, Tito remained interested in space exploration.

Tito was appointed to the Los Angeles Department of Water and Power Board of Commissioners in the 1990s and led the board to support the landmark 1994 state ruling protecting Mono Lake from excessive water diversions by the city.[4]

Space flight

In a project first arranged by MirCorp, Tito was accepted by the Russian Federal Space Agency as a candidate. Tito met criticism from NASA before the launch, primarily from Daniel Goldin, at that time the Administrator of NASA, who considered it inappropriate for a tourist to take a ride into space.[5] [6] When Tito arrived at the Johnson Space Center for additional training on the American sections of the ISS, Robert D. Cabana, NASA manager, sent Tito and his two fellow cosmonauts home, stating "...We will not be able to begin training, because we are not willing to train with Dennis Tito."[7]

Later, through an arrangement with space tourism company Space Adventures, Ltd., Tito joined Soyuz TM-32 on April 28, 2001, spending 7 days, 22 hours, 4 minutes in space and orbiting Earth 128 times.[8] Tito performed several scientific experiments in orbit that he said would be useful for his company and business. Tito paid a reported $20 million for his trip.[9]

Since returning from space, he has testified at the Senate Committee on Commerce, Science & Transportation, Subcommittee on Science, Technology, and Space and the House Committee on Science, Subcommittee on Space & Aeronautics Joint Hearing on "Commercial Human Spaceflight" on July 24, 2003.[10]

See also

- Wilshire Associates
- List of space travelers by name

Further reading

- Klerkx, Greg. Lost in Space, The Fall of NASA. Random House: New York. 2004. ISBN 0-375-42150-5

External links

- Tito the spaceman [11] BBC web site (accessed Feb. 2006)
- Spacefacts biography of Dennis Tito [12]

References

[1] Dennis Tito to Speak at Rensselaer Nov. 14 (https://www.rpi.edu/dept/NewsComm/Campus_News/nov_02/nov_4/tito.html)
[2] LAWRENCE E. DAVANZO APPOINTED PRESIDENT OF WILSHIRE ASSOCIATES (http://www.wilshire.com/Company/PressRoom/PressReleases/Article.html?article=WARelease071910.htm)
[3] DENNIS A. TITO TO RECEIVE PRESTIGIOUS AMERICANISM AWARD FROM BOY SCOUTS OF AMERICA (http://bsa-la.org/about/press_releases/tito2 01-11-2002.html)
[4] Defender of the Trust Award (http://www.monolake.org/mlc/defender)
[5] Goldin, Koptev at Odds on Tito Flight (http://www.space.com/businesstechnology/goldin_tito_010315.html)
[6] NASA Chief Remains Miffed Over Tito Launch: 'Space is Not About Egos' (http://www.space.com/missionlaunches/missions/tito_pm_010428.html)
[7] The forgotten frontier (http://www.sfgate.com/cgi-bin/article.cgi?f=/c/a/2004/01/18/RVGH546K931.DTL)
[8] [http://www.redorbit.com/news/space/65010/first_space_tourist_dennis_tito_to_make_business_visit_to/index.html
[9] Tito the spaceman (http://news.bbc.co.uk/1/hi/sci/tech/1297924.stm) BBC web site (accessed Feb. 2006)
[10] Congressional Testimony "Commercial Human Spaceflight" by Dennis Tito, Wilshire Associates (http://www.spaceref.com/news/viewsr.html?pid=9842)
[11] http://news.bbc.co.uk/1/hi/sci/tech/1297924.stm
[12] http://www.spacefacts.de/bios/astronauts/english/tito_dennis.htm

Mark Shuttleworth

Mark Shuttleworth	

Spaceflight Participant	
Born	18 September 1973 Welkom, South Africa
Other occupation	Entrepreneur
Time in space	9d 21h 25m
Selection	2001
Missions	Soyuz TM-34, Soyuz TM-33
Mission insignia	

Mark Richard Shuttleworth (born 18 September 1973) is a South African entrepreneur who was the second self-funded space tourist.[1] [2] Shuttleworth founded Canonical Ltd. and as of 2010, provides leadership for the Ubuntu operating system. He has a net worth of £150 million ($225 million).[3] He currently lives in the Isle of Man and holds dual citizenship of South Africa and the United Kingdom.

Early life

Shuttleworth was born in Welkom, Free State, South Africa as a son of a surgeon and a nursery school teacher.[4] After attending school at Rondebosch Boys' High School and Diocesan College, Shuttleworth obtained a Business Science degree in Finance and Information Systems at the University of Cape Town. He lived in Smuts Hall, where he was involved in the installation of the first residential Internet connections at the university.

Work

Shuttleworth founded Thawte in 1995, which specialised in digital certificates and Internet security and then sold it to VeriSign in December 1999, earning R 3.5 billion (about US$ 575 million at the time). In September 2000, Shuttleworth formed HBD Venture Capital, a business incubator and venture capital provider. In March 2004 he formed Canonical Ltd., for the promotion and commercial support of free software projects. In December 2009, Shuttleworth stepped down as the C.E.O. of Canonical, Ltd.[5]

Involvement in Linux and FLOSS

In the 1990s, Shuttleworth participated as one of the developers of the Debian operating system.[6]

In 2001 he formed the Shuttleworth Foundation, a non-profit organization dedicated to social innovation which also funds educational, free, and open source software projects in South Africa, such as the Freedom Toaster.

In 2004 he returned to the free software world by funding the development of Ubuntu, a Linux distribution based on Debian, through his company Canonical Ltd.

In 2005 he founded the Ubuntu Foundation and made an initial investment of 10 million dollars. In the Ubuntu project, Shuttleworth is often referred to with the tongue-in-cheek title *Self-Appointed Benevolent Dictator for Life*, abbreviated *SABDFL*.[7] To come up with a list of names of people to hire for the project, Shuttleworth took six months of Debian mailing list archives with him while travelling to Antarctica aboard the icebreaker Kapitan Khlebnikov in early 2004.[8] In September 2005, he purchased a 65% stake of Impi Linux.[9]

Mark Shuttleworth delivering a talk during the Ubuntu Party in Paris in Nov. 2009

On 15 October 2006 it was announced that Mark Shuttleworth became the first patron of KDE, the highest level of sponsorship available.[10]

On 17 December 2009 Mark announced that, effective March of 2010, he would step down as CEO of Canonical to focus energy on product design, partnership and customers. Jane Silber, COO at Canonical since 2004, will take on the job of CEO at Canonical.[11]

Spaceflight

Shuttleworth gained worldwide fame on 25 April 2002 as the second self-funded spaceflight participant. Working with Space Adventures, he launched aboard the Russian Soyuz TM-34 mission, paying approximately US$ 20 million for the voyage. Two days later, the Soyuz spacecraft arrived at the International Space Station, where he spent eight days participating in experiments related to AIDS and genome research. On 5 May 2002, he returned to Earth on Soyuz TM-33. In order to participate on the flight, Shuttleworth had to undergo one year of training and preparation, including seven months spent in Star City, Russia.

Shuttleworth in the International Space Station

While in space he had a radio conversation with Nelson Mandela and a 14 year old South African girl, Michelle Foster, who asked him to marry her. He politely dodged the question, stating that he was "very honoured at the question" before changing the subject.[12] The terminally ill Miss Foster was provided the opportunity to have a conversation with Mark Shuttleworth and Nelson Mandela by the Reach for a Dream foundation.[13] [14]

Transport

He has a private jet, a Bombardier Global Express,[15] which is often referred to as *Canonical One*[16] [17] but is in fact owned through his *HBD Venture Capital* company. The dragon depicted on the side of the plane is "Norman", the HBD Venture Capital mascot.

External links

- Mark Shuttleworth's Homepage [18]
- Spacefacts biography of Mark Shuttleworth [19]
- Interview, Financial Times, January 2006 [20]

References

[1] africaninspace.com (2002). "First African in Space" (http://www.africaninspace.com/home/mission/logs/1/20020610.shtml). HBD. . Retrieved 2007-10-04.

[2] Shuttleworth is the first citizen of an independent African country to go into space. Patrick Baudry, an earlier astronaut, was also born in Africa; however, Baudry's native Cameroon was a French colony at the time of his birth, he is considered a French citizen (although Shuttleworth also had British citizenship at the time of his flight).

[3] "Rich List 2009: Mark Shuttleworth" (http://business.timesonline.co.uk/tol/business/specials/rich_list/rich_list_2009/article6066953.ece). *Sunday Times* (London). 2009-04-26. . Retrieved 2010-03-28. "Results for 2009, Ranking: 362, Worth: £150m"

[4] Vance, Ashlee (2009-01-10). "A Software Populist Who Doesn't Do Windows" (http://www.nytimes.com/2009/01/11/business/11ubuntu.html?pagewanted=all). *New York Times*. . Retrieved 2009-01-12. "charismatic 35-year-old billionaire from South Africa ... son of a surgeon and a kindergarten teacher"

[5] Mark Shuttleworth (2009-12-17). "My new focus at Canonical" (http://www.markshuttleworth.com/archives/295). . Retrieved 2009-12-21.

[6] "debian.org Developers LDAP Search" (http://db.debian.org/). The Debian Project. . Retrieved 2010-04-20. "User Mark Shuttleworth (login "marks", PGP/GPG key id 0xD54F0847)"

[7] "Ubuntu carves niche in Linux landscape" (http://news.com.com/Ubuntu+carves+niche+in+Linux+landscape/2100-7344_3-5886194.html). CNET. .

[8] Linux Format, Jeff Waugh (LXF 87) (http://www.linuxformat.co.uk/waugh.html).

[9] "Shuttleworth bets on ImpiLinux" (http://mybroadband.co.za/nephp/?m=show&id=828). MyADSL. 29 September 2005. . Retrieved 2006-08-28.

[10] "Mark Shuttleworth Becomes the First Patron of KDE" (http://dot.kde.org/2006/10/15/mark-shuttleworth-becomes-first-patron-kde). KDE. 15 October 2006. . Retrieved 2006-10-16.

[11] "Mark Shuttleworth steps down as CEO of Canonical" (http://www.markshuttleworth.com/archives/295). Mark Shuttleworth. 17 December 2009. . Retrieved 2009-12-17.

[12] Space.com, Nelson Mandela Chats with Shuttleworth (http://www.space.com/missionlaunches/mandela_shuttleworth_020502.html), 2002-05-02.

[13] BBC News, Afronaut mourns his 'bride' (http://news.bbc.co.uk/2/hi/africa/2012307.stm), 2002-05-28.

[14] Dispatch online, Mark's biggest fan dies of cancer (http://www.dispatch.co.za/2002/05/28/southafrica/CANCER.HTM), 2002-05-28.

[15] Airliners.net: Bombardier BD-700-1A10 Global Express (http://www.airliners.net/open.file/0805262)

[16] Ubuntu News #16: *Akademy 2006* (https://wiki.ubuntu.com/UbuntuWeeklyNewsletter/Issue16#head-54c204893e3c82b8629568b7372811780fcc50a4)

[17] Ask Slashdot: *Mark Shuttleworth "Canonical One doesn't *actually* belong to Canonical"* (http://interviews.slashdot.org/article.pl?sid=05/04/04/1859255)

[18] http://www.markshuttleworth.com/

[19] http://www.spacefacts.de/bios/international/english/shuttleworth_mark.htm

[20] http://www.benking.co.uk/art/Ubuntu_the_entrepreneur_who_wants_to_give_it_all_away.php

Gregory Olsen

Gregory Olsen	

Spaceflight Participant	
Born	April 20, 1945 (Age: 64) Brooklyn, New York
Other occupation	Entrepreneur
Rank	Spaceflight participant
Time in space	9d 21h 15m
Selection	2005
Missions	Soyuz TMA-7, Soyuz TMA-6
Mission insignia	

Gregory Hammond "Greg" Olsen (born April 20, 1945) is an American entrepreneur and scientist who, in October 2005, became the third private citizen to make a self-funded trip into space with Space Adventures.

Olsen was the co-founder and chairman of Sensors Unlimited Inc., a company developing optoelectronic devices such as sensitive near-infrared (NIR) and shortwave-infrared (SWIR) cameras. One of Sensors Unlimited's major customers is NASA. Currently, Olsen is President of GHO Ventures, LLC, in Princeton, New Jersey, where he manages his angel investments, South African winery, Montana ranch, and performs numerous speaking engagements to encourage children – especially minority and female children – to consider careers in science or engineering. He also is a professor at Rider University where he teaches a physics class.

Early life and education

Olsen, born in Brooklyn, New York, was the son of an IBEW Local 3 electrician. He graduated from Ridgefield Park High School, Ridgefield Park, New Jersey in 1962. After being written off as a failure by teachers due to poor grades in high school, Olsen planned to join the United States Army until he was counseled to try college for several months. Through an IBEW Local 3 scholarship, Olsen attempted college, kept his grades high, and graduated *magna cum laude* with multiple degrees from Fairleigh Dickinson University.[1] He later graduated with a PhD from the University of Virginia.

Olsen admits to little business training and believes that for companies making less than 100 million (the smaller companies as he calls them) that success is based more on "intuition, instinct and hard work." He does credit his success to his graduate science training. "Two of my start-up companies are from the fields I trained in. For instance, my first company EPITAXX (a supplier of optical detectors and receivers for fiber optic telecommunications and cable television networks) relied on my knowledge of physics and material science." Olsen likes to put his money into high-risk start-ups. [2]

Spaceflight details

Flown to the International Space Station (ISS) with Soyuz TMA-7 (launched October 1, 2005, docked October 3) and landed with Soyuz TMA-6 (October 10), Olsen is the third self-funded space tourist to visit the ISS, following Dennis Tito (2001) and Mark Shuttleworth (2002) (all three space tourists flew through Space Adventures, Ltd.). Gregory Olsen has made some comments indicating that he is unhappy with the "space tourist" designation. The following is from National Geographic's coverage "Space Launch - Along for the Ride (2007)": "Greg: The term space tourist implies that you`ll write a check and you go for a joyride. And believe me that is not the case at all. Narrator: Greg worked hard to get this far, training for two years with the Russian Space Agency."

He conducted several experiments in remote sensing and astronomy while aboard the space station. Dr. Olsen is a licensed Amateur Radio radio operator holding FCC callsign KC2ONX and spoke to students via ham radio from space through the ARISS project. In an informal presentation at a New Jersey high school, Dr. Olsen estimated the price of his space excursion at US$20 million.

See also

- List of space travelers by name

Sources

1. Dr. Olsen's presentation at Hopewell Valley Regional High School, Pennington, New Jersey, USA on 26 April 2006.
2. Dr. Olsen's presentation at Montgomery Township High School, Skillman, New Jersey, USA on 20 May 2006
3. Dr. Olsen's presentation at Bear Tavern Elementary school Hopewell Valley Regional school, Titusville New Jersey,USA on 26 November, 2008

External links

- Greg Olsen, Official Web Site [3] GHO Ventures, LLC
- 'Space tourist' blasts off to ISS [4] – BBC, 1 October 2005
- 'Space tourist' arrives at Space Station [5] – BBC, 3 October 2005
- Trek to Space Station Is 'A Dream Come True' [6]
- Spacefacts biography of Gregory Olsen [7]
- Pictures and narrative of the Soyuz TMA-7 launch [8]
- http://space-forum.org/

References

[1] Carroll, Kathleen. "FDU renames engineering school after $5M gift" (http://alacarte.lexisnexis.com/partners/int/google/landingpage.asp?id=14187460&mtid=1&ws=9j0hDk1UboE=&ws_pub=The Record (Bergen County, NJ)&ws_date=May 4, 2006&ws_len=203&ws_lni=4JW7-CC30-TWCN-J1RS-00000-00&ws_title=FDU renames engineering school after&ws_refer=http://news.google.com/archivesearch?um=1&tab=wn&client=firefox-a&ie=UTF-8&oe=utf-8&rls=org.mozilla:en-US:official&q="olsen"+"fairleigh+dickinson"), *The Record (Bergen County)*, May 4, 2006. Accessed June 9, 2007. "About 200 people gathered at Fairleigh Dickinson University on Wednesday to celebrate the university's largest-ever gift and the renaming of one of its schools. Alumnus and trustee Gregory Olsen, the world's third civilian astronaut and the founder of two fiber-optics companies, was greeted with a standing ovation..."

[2] Nagy, Kim; and Stocke, Joy "Reaching for the Stars: An Interview with Greg Olsen - Scientist, Entrepreneur, and Space Traveler" (http://www.wildriverreview.com/spotlight_olsen.php), *Wild River Review*, November 2007. Accessed November 19, 2007.

[3] http://ghoventures.com/bio.aspx

[4] http://news.bbc.co.uk/1/hi/sci/tech/4298814.stm

[5] http://news.bbc.co.uk/2/hi/science/nature/4303596.stm

[6] http://www.spacedaily.com/news/tourism-05zk.html
[7] http://www.spacefacts.de/bios/astronauts/english/olsen_gregory.htm
[8] http://www.htdocs.com/essays/baikonur/

Anousheh Ansari

Anousheh Ansari یراصنا هشونا	

Spaceflight participant	
Born	(aged 43) Mashhad, Iran
Other occupation	Businessperson
Time in space	10d 21h 04m
Selection	2006
Missions	Soyuz TMA-9, Soyuz TMA-8
Mission insignia	

Anousheh Ansari (Persian: یراصنا هشونا, born September 12, 1966 in Mashhad, Iran) is an engineer and the Iranian-American co-founder and chairman of Prodea Systems, Inc and a spaceflight participant with the Russian space program. Her previous business accomplishments include serving as co-founder and CEO of Telecom Technologies, Inc. (TTI). The Ansari family is also the title sponsor of the Ansari X PRIZE. On September 18, 2006, a few days after her 40th birthday, she became the first Iranian in space.[1] Ansari is the fourth overall self-funded space tourist, and the first self-funded woman to fly to the International Space Station. Her memoir, My Dream of Stars, co-written with Homer Hickam, was published by Palgrave-McMillan in 2010.

Early life

Anousheh Ansari born in Mashhad, Iran, Anousheh and her parents moved to Tehran shortly afterward.[2] Anousheh witnessed the Iranian Revolution in 1979. She immigrated to the United States in 1984 as a teenager.[3] Apart from her native Persian, she is fluent in English and French, and acquired a working knowledge of Russian for her spaceflight experience.[4]

She received her Bachelor of Science degree in electrical engineering and computer science at George Mason University and her master's degree at George Washington University.[5] [6] [1]

After graduation, Anousheh began work at MCI, where she met her future husband, Hamid Ansari. They married in 1991.[7] Mrs. Ansari is proud of her Persian roots.[8] [9]

Business career

In 1993, she persuaded her husband, Hamid Ansari, and her brother-in-law, Amir Ansari, to co-found Telecom Technologies, Inc., using their savings and corporate retirement accounts, as a wave of deregulation hit the telecommunications industry. The company was acquired by Sonus Networks, Inc. in 2000 for $550 million in stock. Since the sale, Sonus' stock has fallen from $40 a share to under $5. Ansari and eight other individuals are defendants in a shareholder suit that, among other things, accuses her of insider trading.[10]

Spaceflight

Ansari has expressed that she does not consider herself a "space tourist", and prefers the title of "spaceflight participant".[11]

Ansari holds a plant grown in the Zvezda Service Module of the International Space Station.

Ansari is a member of the X PRIZE Foundation's Vision Circle, as well as its Board of Trustees.[12] Along with her brother-in-law, Amir Ansari, she made a multi-million dollar contribution to the X PRIZE foundation on May 5, 2004, the 43rd anniversary of Alan Shepard's sub-orbital spaceflight. The X PRIZE was officially renamed the *Ansari X PRIZE* in honor of their donation.

The Ansari family investment firm, also named Prodea, has announced a partnership with Space Adventures, Ltd. and the Federal Space Agency of the Russian Federation (FSA) to create a fleet of suborbital spaceflight vehicles (the Space Adventures Explorer) for global commercial use.[13]

Spaceflight

Ansari trained as a backup for Daisuke Enomoto for a Soyuz flight to the International Space Station, through Space Adventures, Ltd.[14] On August 21, 2006, Enomoto was medically disqualified from flying the Soyuz TMA-9 mission that was due to launch the following month. The next day Ansari was elevated to the prime crew.[15] [16]

Asked what she hoped to achieve on her spaceflight, Ansari said, "I hope to inspire everyone—especially young people, women, and young girls all over the world, and in Middle Eastern countries that do not provide women with the same opportunities as men—to not give up their dreams and to pursue them...It may seem impossible to them at times. But I believe they can realize their dreams if they keep it in their hearts, nurture it, and look for opportunities and make those opportunities happen."[17] The day before her departure, she was interviewed on Iran national television for the astronomy show *Night's Sky*.[18] The hosts wished her success and thanked her on behalf of Iranians. Ansari in return, thanked them.[19]

Ansari lifted off on the Soyuz TMA-9 mission with commander Mikhail Tyurin (RSA) and flight engineer Michael Lopez-Alegria (NASA) at 04:59 (UTC) on Monday September 18, 2006 from Baikonur, Kazakhstan. Ansari became the fourth (and first female) space tourist. Her contract did not allow for disclosure of the amount paid, but previous space tourists have paid in excess of $20 million USD. The space craft docked with the International Space Station (ISS) on Wednesday September 20, 2006, at 05:21 (UTC).[20] [21] Ansari landed safely aboard Soyuz TMA-8 on September 29, 2006 at 01:13 UTC on the steppes of Kazakhstan (90 kilometers north of Arkalyk) with U.S. astronaut Jeffrey Williams and Russian cosmonaut Pavel Vinogradov.[22] She was given red roses from an unidentified official, and a surprise kiss from her husband, Hamid.[23] [24] Rescuers moved them to Kustanai for welcome ceremony with helicopters.[25] [26]

During her eight day stay on board the International Space Station, Ansari agreed to perform a series of experiments on behalf of the European Space Agency. She conducted four experiments,[27] including:

- Researching the mechanisms behind anemia.

- How changes in muscles influence lower back pain.
- Consequences of space radiation on ISS crew members and different species of microbes that have made a home for themselves on the space station.

She also became the first person to publish a weblog from space.

Iranian flag controversy

Ansari intended to wear the U.S. flag on her spacesuit alongside with a politically-neutral version of the Iranian flag, i.e. the simple 3-color flag with no government-specific emblem, to honor the two countries that have contributed to her life.[28] A few U.S.-based media wrongly speculated that she was intending to wear the version of the Iranian flag that predated the 1979 Islamic revolution in Iran.[29]

At the insistence of the NASA and Russian officials, she did not wear the Iranian flag officially, but wore the Iranian flag colors instead and kept the Iranian flag on her official flight patch.[30] She and her husband said no political message was intended, despite the increasing tensions with United States and Iran relations, which had dominated world headlines in the weeks leading up to her launch. She noted that she had "plans to devote her mission to expanding a global consciousness she expected would be seeded with her first look at Earth from space".

Anousheh Ansari's official space flight patch, featuring both the flags of Iran (lower right) and the U.S. (lower left), the ISS icon, and the map of Iran marked in dark green on the Earth's surface.

Reactions to Ansari's flight

Crewmates

Michael Lopez-Alegria, the Spanish-born NASA astronaut who flew on the Russian Soyuz spacecraft on the return flight with Ansari, expressed his doubts to reporters before the flight: "I'm not a big fan personally of having those guys go visit the space station because I think the space station is still a place that is under construction, and not quite operational. I don't think it's ideal."[31]

Lopez-Alegria later stated that he was skeptical of private tourists a few years ago, but now believes it is essential to the survival of the Russian space program which is important to the U.S. space program: "If that's the correct solution... then not only is it good from the standpoint of supporting the Russian space program, but it's good for us as well," he said. Ansari's presence in space "is a great dream and a great hope not just for our country but for countries all around the world."[32]

The same Associated Press story also quoted Mikhail Tyurin describing Ansari as "very professional" and said he felt like they had worked together for a decade.

Reactions in Iran

Reactions to the flight in Ansari's native Iran were generally mixed. It was given significant coverage by state television. IRIB aired a live 1 hour interview with Anousheh in Aseman-e-Shab (Night Sky) live show. Anousheh was praised by newspapers such as *Hambastegi* and *Jam-e-Jam Daily*, which published daily columns detailing the journey. The astronomy magazine *NOJUM* also published an exclusive interview of Pouria Nazemi with Anousheh before her trip, in which she discussed her vision for commercial spaceflight. *NOJUM* also organized and held gatherings when the ISS passed over Iran's cities. Shahram Yazdanpanah, made a special part about Anousheh's trip to space at Persian "Space Science [33]" website and covered all the news of trip.

Crew of Soyuz TMA-9: Astronaut Michael E. Lopez-Alegria (left), Anousheh Ansari (middle) and cosmonaut Mikhail Tyurin at the Cosmonaut Hotel in Baikonur, Kazakhstan on Sept. 5, 2006

Interviews

On September 22, 2006, she told reporters that she has no regrets and said "I am having a wonderful time here. It's been more than what I expected, and I am enjoying every single second of it. The entire experience has been wonderful up here."[34]

Honors and awards

Ansari has received multiple honors, including the George Mason University Entrepreneurial Excellence Award, the George Washington University Distinguished Alumni Achievement Award, and the Ernst & Young Entrepreneur of the Year Award for the Southwest Region. While under her leadership, Telecom Technologies, Inc. earned recognition as one of Inc. magazine's 500 fastest-growing companies and one of Deloitte & Touche's Fast 500 technology companies. She was listed in Fortune magazine's "40 under 40" list in 2001 and honored by Working Woman magazine as the winner of the 2000 National Entrepreneurial Excellence award. On 2009 she was received the first Symons Innovator Award [35] given annually by NCWIT [36] to honor successful women entrepreneurs in technology.

The Ansari family was recently honored with an Orbit Award by the National Space Society and Space Tourism Society for underwriting the Ansari X PRIZE.

Anousheh Ansari participated as a speaker at the 2010 Honeywell Leadership Academy at Space Camp in Huntsville Alabama with Homer Hickam. [37]

See also

- Iranian women
- List of famous Persian women
- List of International Space Station visitors
- List of Muslim astronauts

External links

- Personal website [38]
- Anousheh Ansari Space Blog [39]
- Space Explorer, Anousheh's space album on Flickr [40]
- Spacefacts biography of Anousheh Ansari [41]

- Space Tourists, a documentary film from Christian Frei with Anousheh Ansari [42]
- Entrepreneurial Heroes [43] - podcast interview with NCWIT
- Anousheh Ansari in SpaceScience.ir (in persian) [33]
- Interview with Anousheh Ansari [44]
- A 'passion' for space travel BBC/Click Video Interview with Anousheh Ansari [45]

References

[1] Andrew Buncombe (2006-09-18). "Pride in space as Iran cheers first Muslim's journey to the stars".

[2] "Alumni news" (http://ece.gmu.edu/alumni_news.htm). George Mason University. 2001. . Retrieved 2006-09-22.

[3] "Female space tourist blasts off" (http://www.cnn.com/2006/TECH/space/09/18/space.female.tourist/index.html). CNN. 2006-09-18. . Retrieved 2006-09-18.

[4] WARREN E. LEARY, *She Dreamed of the Stars; Now She'll Almost Touch Them* (http://www.nytimes.com/2006/09/12/science/space/12tourist.html?_r=1&ref=science&oref=slogin), NY Times , September 12, 2006

[5] Emily Yaghmour. "Failure Was Never an Option for Mason Alumna" (http://www.gmu.edu/alumni/spirit/00fall/ansari.html). *George Mason University alumni newsletter*. . Retrieved 2006-08-27.

[6] GINA SUNSERI, *First Female Space Tourist Takes Off* (http://abcnews.go.com/Technology/story?id=2455201&page=1) ABC News, Sept. 18, 2006

[7] "First female space tourist poised for launch". CNN .com. 2006-09-15.

[8] "Anousheh Ansari Proud of Being Iranian Muslim ", 2008-12-10. Fars news agency. excerpt: "I have always advertised my Persian roots and I'm proud to do so. I always start my story by telling everyone that I was born in Iran and show the map of Iran with my place of birth, Mashhad." (http://english.farsnews.com/newstext.php?nn=8709201466) accessed May 26, 2009.

[9] "Ovi Magazine", Interview with Anousheh Ansari: Iran's first space explorer by Kourosh Ziabari, 2008-12-18. Accessed May 26 2009, (http://www.ovimagazine.com/art/3857).

[10] Koeng, David (2006-09-16). "Iranian-Born Woman to Be Space Tourist". Lycos News.

[11] "Space for Business Newsletter article from February 2007: "I am NOT a tourist"" (http://www.esa.int/esaHS/SEMD58BE8YE_index_0.html). . February 2007

[12] "Board of Trustees" (http://www.xprizefoundation.com/about_us/board.asp). X PRIZE Foundation. . Retrieved 2006-09-11.

[13] "Space Tourism Pioneers, Space Adventures, and the Ansari X PRIZE Title Sponsors to Provide First Suborbital Spaceflight Tourism Vehicles" (http://www.prnewswire.com/cgi-bin/stories.pl?ACCT=104&STORY=/www/story/02-16-2006/0004283899&EDATE=). *PR Newswire*. . Retrieved 2006-08-27.

[14] "Iranian Woman Blazes Trail Into Space" (http://www.space-travel.com/reports/Iranian_Woman_Blazes_Trail_Into_Space_999.html). *Spacedaily.com*. . Retrieved 2006-08-27.

[15] "Iranian-born American approved to replace Japanese space tourist" (http://en.rian.ru/world/20060822/53002090.html). *Novosti*. . Retrieved 2006-08-22.

[16] "NASA Expedition 14 Overview" (http://www.nasa.gov/mission_pages/station/expeditions/expedition14/exp14_advancer.html). .

[17] "Interview with Anousheh Ansari, the First Female Space Tourist" (http://www.space.com/missionlaunches/060915_ansari_qna.html). . Retrieved 2006-09-19.

[18] "Up, Up, and Away!" (http://www.iranian.com/Kasraie/2004/October/Ansari/index.html). .

[19] "Full Coverage of Anousheh's journey to space" (http://www.spacescience.ir/eng-Anousheh_Ansari.htm). .

[20] ""Lift-off for woman space tourist "" (http://news.bbc.co.uk/2/hi/science/nature/5355022.stm). BBC News Online. 2006-09-18. . Retrieved 2006-09-22.

[21] ""Space tourist, new crew board ISS"" (http://news.bbc.co.uk/2/hi/science/nature/5362944.stm). BBC News Online. 2006-09-20. . Retrieved 2006-09-22.

[22] "INTERVIEW: From space, a new view of an Iranian" (http://asia.news.yahoo.com/060916/3/2pzp7.html). *Yahoo*. . Retrieved 2006-09-18.

[23] ""Space Station Crew Back on Earth"" (http://www.nasa.gov/mission_pages/station/expeditions/expedition13/exp13_undocking.html). NASA. 2006-09-28. . Retrieved 2006-09-30.

[24] ""Space tourist in Earth touchdown"" (http://news.bbc.co.uk/2/hi/science/nature/5390902.stm). BBC News Online. 2006-09-29. . Retrieved 2006-09-30.

[25] ""American female space tourist returns"". The Detroit News. 2006-09-29.

[26] ""Space triumph prompts new line in underwear"" (http://icwales.icnetwork.co.uk/0100news/0700world/tm_objectid=17841909&method=full&siteid=50082&headline=space-triumph-prompts-new-line-in-underwear-name_page.html). Western MaillicWales, The Western Mail. 2006-09-29. . Retrieved 2006-09-30.

[27] "IESA experiments with spaceflight participant Ansari to ISS" (http://www.esa.int/esaCP/SEMQRH7LURE_index_0.html). . Retrieved 2006-09-22.

[28] Ender, West (2006-08-28). "Iranian flag in Space" (http://www.persianstudents.org/archives/002273.html). Persian Students in the UK Weblog. . Retrieved 2007-01-07.

[29] Maher, Heather (2006-09-15). "U.S.: Iranian-American To Be First Female Civilian In Space" (http://www.rferl.org/featuresarticle/2006/09/141fd0f7-8cf3-45e0-83e4-d534dc4ea152.html). Radio Free Europe. . Retrieved 2006-09-17.
[30] Slater, Shelly (2006-09-14). "Local space tourist's Iran patch spurs dispute" (http://www.wfaa.com/sharedcontent/dws/wfaa/latestnews/stories/wfaa060914_mo_womanspace.d4fdda8.html). . Retrieved 2006-09-17.
[31] Than, Ker (2006-09-16). "First Female Space Tourist, Next ISS Crew Set to Launch" (http://www.space.com/missionlaunches/060916_exp14_prelaunch.html). space.com. . Retrieved 2006-09-18.
[32] Eckel, Mike (2006-09-18). "First female space tourist blasts off". Yahoo News.
[33] http://www.spacescience.ir/Anousheh_Ansari.htm
[34] "First female space tourist savoring 'every single second' in orbit". AP. 2006-09-22.
[35] http://www.ncwit.org/work.awards.innovator.html
[36] http://www.ncwit.org
[37] "2009HoneywellLeadershipGuide" (http://www.spacecamp.com/forms/2009HoneywellLeadershipGuide.pdf). .
[38] http://www.anoushehansari.com/
[39] http://spaceblog.xprize.org/
[40] http://www.flickr.com/photos/spaceexplorer/
[41] http://www.spacefacts.de/bios/astronauts/english/ansari_anousheh.htm
[42] http://www.space-tourists-film.com/
[43] http://www.ncwit.org/resources.profiles.php?source=&action=display&interview_id=256&year=2007&month=11
[44] http://ovimagazine.com/art/3857
[45] http://news.bbc.co.uk/2/hi/programmes/click_online/archive/7968022.stm

Charles Simonyi

Charles Simonyi	

Born	September 10, 1948 Budapest, Hungary
Nationality	Hungarian
Occupation	Software architect
Known for	Hungarian notation, space tourism
Notable works	Microsoft Office

Charles Simonyi (Hungarian: *Simonyi Károly*; born September 10, 1948) is a Hungarian-American computer software executive who, as head of Microsoft's application software group, oversaw the creation of Microsoft's flagship office applications.[1] [2] He now heads his own company, *Intentional Software*,[1] with the aim of developing and marketing his concept of intentional programming. In April 2007, aboard Soyuz TMA-10, he became the fifth space tourist and the second Hungarian in space. In March 2009, aboard Soyuz TMA-14, he made a second trip to the International Space Station. His estimated net worth is US$1 billion.[3]

Biography

Early life in Hungary

Simonyi was born in Budapest, Hungary, the son of Simonyi Károly, a professor of electrical engineering[4] at Technical University of Budapest. While in high school he worked part-time as a night watchman at a computer laboratory, overseeing a large Soviet Ural II[5] mainframe. He took an interest in computing[6] and learned to program from one of the laboratory's engineers. By the time he left school, he had learned to develop compilers and sold one of these to a government department. He presented a demonstration of his compiler to the members of a Danish computer trade delegation.[6]

Denmark and USA

He was hired by Denmark's A/S Regnecentralen in 1966 and moved to the United States in 1968 to attend the University of California, Berkeley, where he earned his B.S. in Engineering Mathematics & Statistics in 1972.

Simonyi then went to Stanford University[1] for graduate studies and was hired by Xerox PARC[1] during its most productive period, working alongside luminaries such as Alan Kay, Butler Lampson and Robert Metcalfe on the development of the *Xerox Alto*, the first personal computer. He and Lampson developed *Bravo*, the first WYSIWYG document preparation program, which became operational in 1974. He received his Ph.D. in computer science from Stanford in 1977 with a dissertation on a software project management technique called "metaprogramming". This approach sought to defeat Brooks' law by requiring all programmers to communicate through the manager rather than directly. Simonyi remained at PARC until 1981.

Microsoft

In 1981, at Metcalfe's suggestion, he applied directly to Bill Gates for a job at Microsoft.[6] At the firm, Simonyi oversaw the development of what became its most profitable products, Word and Excel, as well as Excel's predecessor Multiplan. With Multiplan, Simonyi pursued a strategy called the "revenue bomb", whereby the product ran on a virtual machine that was ported to each platform. The resulting application was highly portable, although Simonyi did not foresee the rapid adoption of MS-DOS that made such efforts less important.[7] Simonyi introduced the techniques of object-oriented programming that he had learned at Xerox to Microsoft. He developed the Hungarian notation convention[8] for naming variables. Originally these standards were part of his doctoral thesis.[8] The Hungarian notation has been widely used inside Microsoft.[9]

Own company

Simonyi remained at Microsoft during its rapid rise in the software industry, becoming one of its highest-ranking developers. He left abruptly in 2002 to co-found, with business partner Gregor Kiczales, a company called Intentional Software.[10] This company markets the intentional programming concepts Simonyi developed at Microsoft Research. In this approach to software, a programmer first builds a toolbox specific to a given problem domain (such as life insurance). Domain experts, aided by the programmer, then describe the program's intended behavior in a What You See Is What You Get (WYSIWYG)-like manner. An automated system uses the program description and the toolbox to generate the final program. Successive changes are only done at the WYSIWYG level.[11]

In 2004, Simonyi received the Wharton Infosys Business Transformation Award for the industry-wide impact of his innovative work in information technology.

Philanthropy

Simonyi has been an active philanthropist. In 1995 he established an endowed chair, the Simonyi Professorship of the Public Understanding of Science at Oxford University, first held by the now retired Richard Dawkins. He also established a Charles Simonyi Professor for Innovation in Teaching endowed chair at Stanford University. In January 2004, Simonyi created the $50 million Charles Simonyi Fund for Arts and Sciences, through which Simonyi plans to support Seattle-area arts, science, and educational programs. Initial grant recipients include the Seattle Symphony ($10 million), and the Seattle Public Library ($3 million). In 2005, the Fund donated $25 million to the Institute for Advanced Study in Princeton, New Jersey. In January, 2008 the Simonyi Fund and Bill Gates pledged $20 million and $10 million respectively to the Large Synoptic Survey Telescope.

Personal life

As of February 2008, Simonyi had been dating Martha Stewart for 15 years.[12] In 2008, however, the two separated. On August 8, 2008, Simonyi became engaged to a Swedish millionaire's daughter, Lisa Persdotter, who is 32 years his junior. On November 22, 2008, the couple married in a private ceremony in Gothenburg, Sweden attended by their closest friends, among them Bill Gates.[13] [14]

Simonyi spends six months a year[15] on his custom-built 233 ft (71 m) long super yacht named *Skat*.[15]

Simonyi's residence in Medina, Washington, "Villa Simonyi", is a modern house designed by architect Wendell Lovett, where Simonyi displays his collection of paintings by Roy Lichtenstein and Victor Vasarely.[16] [17] [18]

Space tourist

In early 2006, Simonyi expressed interest in becoming a space tourist and signed agreements with the space tourism company, Space Adventures, Ltd., for a ten-day mission to the International Space Station (ISS).

In August 2006, he passed a pre-qualification medical exam by the Russian Federal Space Agency, called the State Medical Commission (GMK). He started training at Star City in September 2006.[2]

He launched on April 7, 2007 (GMT),[19] on board Soyuz TMA-10. He shared a ride with two Russian cosmonauts to the International Space Station,[5] and returned aboard Soyuz TMA-9, scheduled to depart from the ISS on April 20, 2007.

Upon arrival to the ISS on April 9, 2007 Simonyi said, "It is amazing how it appears from the blackness of the sky. It was very, very dramatic. It was like a big stage set, a fantastic production of some incredible opera or modern play. That's what I was referring to when I said I was blown away."[20]

Simonyi's expected return on April 20 was delayed by one day due to 'boggy ground'. He returned to Earth on April 21 along with an American astronaut and a Russian cosmonaut.[21]

In October 2008, he booked for a second trip to the ISS through Space Adventures on board Soyuz TMA-14.[22] On March 26, 2009 he returned to space aboard Soyuz TMA-14.[23] He returned to Earth on board Soyuz TMA-13. Along with Soyuz Commander Yuri Lonchakov and Michael Fincke he landed in Kazakhstan on April 8, 2009.[24]

Radio communication while aboard ISS

Charles Simonyi is a licensed amateur radio operator with the call sign KE7KDP, and planned to contact a number of schools while on his flight on the International Space Station utilizing amateur radio for the communication with those schools. On April 11, 2007 the American Radio Relay League reported that Simonyi was already making ham radio contacts from space.[25]

One of the schools Simonyi contacted was Cedar Point Elementary in Bristow, Virginia. A telebridge conversation was held on Tuesday, April 17, 2007. Onboard with him were Oleg Kotov and Fyodor Yurchikhin.[26] On March 30, 2009 he held a phone conversation with students at the Girls' Middle School in Mountain View, CA, USA in which he said that one of the most surprising things about traveling to space was that upon returning to earth the air feels very thick, very heavy, like "breathing Pepto-Bismol." He also stated that talking with the students from the school "made his day."

Simonyi used his Hungarian call sign HA5SIK[27] when he contacted 25 radio amateurs from Hungary in a record attempt on April 12. He contacted former and current students of Tivadar Puskás Polytechnic, Budapest on April 13.[27]

External links

- NASA biography of Charles Simonyi [28]PDF (20.4 KiB)
- Charles in Space [15]
- Charles Simonyi, the benefactor of the Simonyi Professorship at the University of Oxford [29]
- Charles Simonyi Fund for Arts and Sciences [30]
- The Rosetta code: archaeology of the Gier-Algol compiler program (an Intentional Software presentation by Simonyi) [31]
- Forbes interview, 23 December 2002 [32]

References

[1] Lohr, Steve (2002-09-17). "A Microsoft Pioneer Leaves to Strike Out on His Own" (http://query.nytimes.com/gst/fullpage.html?res=9C0DE2DA1730F934A2575AC0A9649C8B63&scp=1&sq=Charles Simonyi&st=cse). *The New York Times*. . Retrieved 2008-05-21.

[2] "'Nerd' outlines space ambitions" (http://news.bbc.co.uk/2/hi/science/nature/6088460.stm). *BBC News*. 2006-10-26. . Retrieved 2008-05-21.

[3] "The World's Billionaires — No.891 Charles Simonyi" (http://www.forbes.com/lists/2007/10/07billionaires_Charles-Simonyi_VMOW.html). *Forbes*. 2007. . Retrieved 2008-05-21.

[4] Peter Weibel; Ludwig Múzeum (Budapest, Hungary), Neue Galerie am Landesmuseum Joanneum, Museum of Modern Art, Antwerp (2005). *Beyond Art: A Third Culture : a Comparative Study in Cultures, Art, and Science in 20th Century Austria and Hungary* (http://books.google.com/books?id=xkk6U42Zl_sC&pg=PA351). Springer. p. 351. ISBN 9783211245620. .

[5] The Associated Press (2007-04-09). "Soyuz capsule carrying U.S. billionaire Charles Simonyi docks at space station" (http://www.iht.com/articles/ap/2007/04/09/europe/EU-GEN-Russia-Space.php). *International Herald Tribune*. The New York Times Company. . Retrieved 2008-07-31.

[6] Alesso, H. Peter; Craig F. Smith, James Burke (2007). "Charles Simonyi" (http://books.google.com/books?id=DxGyOaAyd6gC&pg=PA92#PPA93,M1). *Connections: Patterns of Discovery*. Wiley-IEEE. p. 93. ISBN 9780470118818. .

[7] Robert Mitchell (2002-12-09). "Microsoft Days: Transforming the Desktop" (http://www.computerworld.com/softwaretopics/software/appdev/story/0,10801,76413,00.html). *Computerworld*. . Retrieved 2008-05-21.

[8] Todd Barron (2001). *Multiplayer Game Programming* (http://books.google.com/books?id=A9Z69ql3HrsC&pg=PA220). Thomson Course Technology. p. 220. ISBN 9780761532989. .

[9] "Hungarian Notation" (http://msdn.microsoft.com/en-us/library/aa260976(VS.60).aspx). *Microsoft Developer Network*. Microsoft Corporation. . Retrieved 2008-07-31.

[10] "CHARLES SIMONYI: President and CEO" (http://www.intentionalsoftware.com/company/management.html). *Intentional Software Corporation*. . Retrieved 2008-05-21.

[11] "Anything You Can Do, I Can Do Meta" (http://www.technologyreview.com/Infotech/18047/). *Technology Review*. January 2007. . Retrieved 2008-05-21.

[12] "Martha Stewart Still Stewing Over Donald Trump" (http://www.people.com/people/article/0,,20171480,00.html). *People*. 2008-01-12. . Retrieved 2008-05-21.

[13] Bergfeldt, Carina; Camilla Sundell (November 22, 2008). "Gates vilda festnatt i Göteborg" (http://www.aftonbladet.se/nyheter/article3830531.ab) (in Swedish). *Aftonbladet*. . Retrieved 2009-10-24. (Gates' wild party night in Gothenburg)

[14] Lindwall, Johan T (8 september 2008). "Lisa Persdotter, 28, gifter sig med miljardär" (http://www.gt.se/1.1292042) (in Swedish). *Göteborgs-Tidningen*. . Retrieved 2009-10-25. (Lisa Persdotter, 28, marries billionaire)

[15] Blakeley, Kiri (2006-03-27). "Bigger Than Yours" (http://www.forbes.com/forbes/2006/0327/160.html). *Forbes*. . Retrieved 2009-04-09.

[16] "College of Architecture and Urban Planning Newsletter" (http://www.caup.washington.edu/support/newsletter/2003_summer.pdf) (PDF). Seattle, Washington: University of Washington. Summer 2003. p. 6. . Retrieved 2009-10-24. "Villa Simonyi, the sprawling Medina home Architecture Professor Emeritus Wendell Lovett designed for Charles Simonyi, was the subject of an article in the May 18 edition of Pacific Northwest Magazine."

[17] Lohr, Steve (2001). *Go To: The Story of the Math Majors, Bridge Players, Engineers, Chess Wizards, Maverick Scientists and Iconoclasts - the Programmers who created the software revolution* (http://books.google.com/books?id=XfPLVx6qS_cC&pg=PA1&lpg=PA1&dq=Simonyi+Roy+Lichtenstein+and+Victor+Vasarely&source=bl&ots=U2dMt-qaMI&sig=DlOck-RJ6DCR8RMiNtmEAjHrLz4&hl=en&ei=7b7jSqaJOZCZ8AbUruCHBw&sa=X&oi=book_result&ct=result&resnum=7&ved=0CBwQ6AEwBg#v=onepage&q=Simonyi Roy Lichtenstein and Victor Vasarely&f=false). New York, NY: Basic Books. p. 1. ISBN 0-465-04226-0. .

[18] Seven, Richard (May 16, 2003). "Fresh Angles: The beauty is in the geometric interplay of design and function" (http://community.seattletimes.nwsource.com/archive/?date=20030516&slug=pfit18). *Pacific Northwest Magazine*. . Retrieved 2009-10-24. "Charles Simonyi, a driving force behind the development of some of Microsoft's most successful software, lives in a sprawling Medina home built over the years in sections as he acquired property from neighbors on both sides.

He worked closely with longtime Seattle architect Wendell Lovett to fit idiosyncratic parts into a whole. It is inspired, in part, by one of Simonyi's favorite artists — fellow Hungarian Victor Vasarely, whose optical images have influenced both computer science and architecture."

[19] "American space 'nerd' blasts off" (http://news.bbc.co.uk/1/hi/sci/tech/6529149.stm). *BBC News*. 2007-04-07. . Retrieved 2008-05-22.

[20] "Tourist Talks From Space" (http://news.sky.com/skynews/article/0,,30200-1260128,00.html). *Sky News*. 2007-04-11. . Retrieved 2008-05-22.

[21] "Space tourist makes safe return" (http://news.bbc.co.uk/2/hi/science/nature/6578835.stm). *BBC News*. 2007-04-21. . Retrieved 2008-05-22.

[22] "Software billionaire books 2nd trip to space" (http://technology.sympatico.msn.cbc.ca/abc/news/ContentPosting.aspx?isfa=1&feedname=CBC-TECH-SCIENCE-V3&showbyline=True&newsitemid=simonyi-2ndtrip). *CBC News* (Canadian Broadcasting Corporation).

2008-10-02. . Retrieved 2008-10-02.

[23] Harwood, William (2009-03-26). "Russian Rocket Carries Space Tourist" (http://www.nytimes.com/2009/03/26/science/space/27soyuz.html?_r=1&hp). *The New York Times*. . Retrieved 2009-03-26.

[24] "Space Tourist Returns from Space Station" (http://www.nytimes.com/aponline/2009/04/08/world/AP-EU-Russia-Space.html?_r=1&hpw). *The New York Times*. The Associated Press. 2009-04-08. .

[25] "Civilian Space Traveler Already Making Ham Radio Contacts from Space" (http://www.arrl.org/news/stories/2007/04/11/100/). *American Radio Relay League*. 2007-04-11. . Retrieved 2008-05-22.

[26] Catchpole, John E. (2008). *The International Space Station: Building for the Future* (http://books.google.com/books?id=VsTdriusftgC&pg=PA274&lpg=PA274&dq=Simonyi+Oleg+Kotov+and+Fyodor+Yurchikhin&source=bl&ots=pFaaPSLb-C&sig=KymXATNE3b_2VFUbFi1TOYXYhT8&hl=en&ei=adrjSoHLBo7l8Ab-5biIBw&sa=X&oi=book_result&ct=result&resnum=2&ved=0CAoQ6AEwAQ#v=onepage&q=Simonyi Oleg Kotov and Fyodor Yurchikhin&f=false). Springer-Praxis books in space exploration. Chichester, UK: Praxis Publishing. pp. 274–275. ISBN 978-0-387-78144-0. .

[27] "Rádiókapcsolat Charles Simonyival (Radio Connection with Charles Simonyi)" (http://www.termeszetvilaga.hu/radiokapcsolat_text.html#kozlemeny_apr14_1200) (in Hungarian). *Természet Világa*. 2007-04-12. .

[28] http://www.nasa.gov/pdf/163534main_simonyi.pdf

[29] http://www.simonyi.ox.ac.uk/simonyi/

[30] http://www.simonyifund.org/

[31] http://www.datamuseum.dk/site_dk/20040213/cs/index.html

[32] http://www.forbes.com/forbes/2002/1223/342.html

Richard Garriott

Richard Garriott	

Born	July 4, 1961 Cambridge, England
Nationality	British / American
Occupation	Game designer

Richard Allen Garriott (born July 4, 1961) is a British-American video game developer and entrepreneur. He is also known as his alter egos **Lord British** in *Ultima* and **General British** in *Tabula Rasa*. A figure in the video game industry, Garriott was originally a game designer and programmer and now engages in various aspects of computer game development and business.

On October 12, 2008, Garriott launched aboard Soyuz TMA-13 to the International Space Station as a self-funded tourist[1] [2], returning 12 days later aboard Soyuz TMA-12.

Biography

Early life

Garriott was born in Cambridge, England,[3] and raised in Nassau Bay, Texas. He is the son of scientist Owen K. Garriott, who became an astronaut and flew with Skylab 3 and Space Shuttle mission STS-9. At Clear Creek High School, he took an interest in computers and began self-directed courses in programming in which he created fantasy computer games. In the summer of 1977, Richard's parents sent him to the University of Oklahoma for a seven-week computer camp. Shortly after he arrived, some of the other boys attending the camp introduced themselves. When Garriott replied to their greeting of "Hi" with "Hello" they decided he sounded like he was from Britain, and gave him the nickname "British". Garriott uses the name to this day for his various gaming characters, including Ultima character Lord British and Tabula Rasa character General British,[4] despite the fact that Garriott's parents moved to Texas when he was a baby and — despite his nickname and birthplace — his accent is American rather than British.[5]

Programming career

Richard continued programming many games, often offering them to friends for free. He produced his first published game, *Akalabeth* (*DND28b*[6]), in the summer of 1980 while working at a ComputerLand retail store. Garriott earned more than enough money from *Akalabeth* to pay to continue his education. In the fall, he entered the University of Texas at Austin, joined the school's fencing team and later joined the Society for Creative Anachronism.

In the early 1980s, Garriott developed the *Ultima* computer game series (sequels after the first were numbered, such as *Ultima II*, *Ultima III* and so on). Originally programmed for the Apple II, the first was published by California Pacific Computers, and sold in Ziploc plastic bags to interested parties. The second part was published by Sierra On-Line. By the time he developed his third installment, the games had such a large following that Garriott (along with his brother, Robert, and father and others) established Origin Systems, their own video game publisher, to handle the publishing and distribution of his title, now available on several platforms.

Garriott sold Origin to Electronic Arts in September 1992. In 1997, he coined the term Massively Multiplayer Online Role-Playing Game, MMORPG, giving a new identity to the nascent genre previously known as graphical MUDs.[7] In 1999 and 2000, EA canceled all of Origin's new development projects, including *Privateer Online*, and *Harry Potter Online*.[8] [9] In the midst of these events, Garriott resigned from the company and returned to the industry by forming Destination Games in April 2000 with his brother and Starr Long (the producer of *Ultima Online*). Once Garriott's non-compete agreement with EA expired a year later, Destination partnered with NCsoft where he acted as a producer and designer of MMORPGs. After that, he became the current CEO of NCsoft Austin, also known as NC Interactive.

In an open letter on the *Tabula Rasa* website posted November 11, 2008, Garriott announced his plans to leave NCsoft to pursue new interests sparked by his spaceflight experiences. Later, however, Garriott claimed that the letter was forged as a means of forcing him out of his position and that he had had no intention of leaving.[10] On the 24th of November NCsoft announced they planned to end the live service of *Tabula Rasa*. The servers shut down on February 28, 2009, after a period of free play from January 10 onward for existing account holders.[11]

Spaceflight

On September 28, 2007, Space Adventures announced that Garriott would fly to the International Space Station in October 2008 as the sixth space tourist, reportedly paying $30 million USD.[1] [12] On October 12, 2008, Garriott became the second second-generation space traveler and the first offspring of an American astronaut to go into space,[1] [13] [14] and the second person to wear the British Union flag in space.[15] The Soyuz docked with the station on October 14. Garriott placed a geocache while aboard the ISS.[16] He returned to Earth on October 24 aboard the Soyuz TMA-12 with two members of Expedition 17, cosmonauts Oleg Kononenko and Sergei Volkov. Garriott is the second second-generation space traveler, after Volkov.[17] [18]

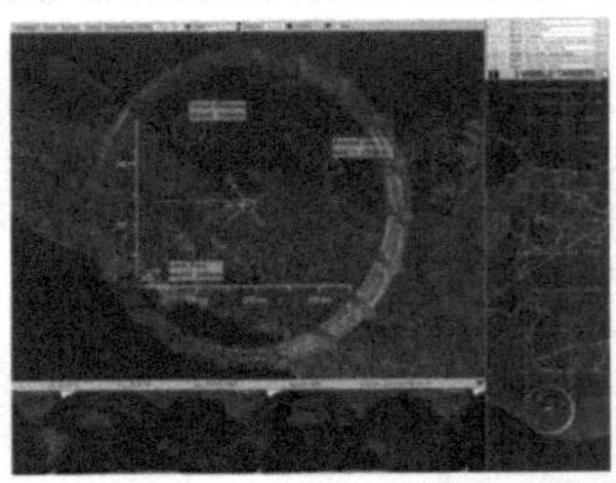

Screen capture from Windows on Earth, used by Garriott on ISS to identify targets for Earth photography. (Coast of Peru)

During his spaceflight, Garriott took part in several education outreach efforts. He is an Amateur Radio Operator (callsign W5KWQ), and during his stay on the International Space Station (ISS), communicated with students and other Amateur Radio operators using Amateur Radio.[19] Garriott also transmitted photographs using the ARISS SSTV [20] amateur radio system. Garriott also worked with the Windows on Earth project, which provides an interactive, virtual view of Earth as seen from the ISS.[21] Garriott used Windows on Earth software to assist in the selection of locations on Earth to photograph, and the public were able to use the same online tool to track the ISS and see the view Richard was experiencing out an ISS window. Richard's photographs, along with images taken by his astronaut father Owen Garriott in 1973, will be available to the public through Windows on Earth, adding a personal element to studies of Earth and how Earth has changed over time.[21]

Also, Tracy Hickman wrote a screenplay for Garriott, for the first science-fiction film shot in space, *Apogee of Fear*.[22]

On October 24, Russian cosmonauts of ISS Crew 17, Sergei Volkov and Oleg Kononenko, along with space tourist Richard Garriott, aboard Soyuz TMA-12 capsule, landed safely (ideal) at 9:37 a.m. (0336GMT, 07:36 Moscow time), 55 miles north of Arkalyk, Kazakhstan. They were lifted to the Kazakhstan Baikonur space center by helicopter, and then flew to Zvezdny Gorodok (Star City), Moscow Region.[23] [24] [25] [26]

On June 3, 2009, the *New York Daily News* announced that Garriott would officiate at the first wedding to be held in zero gravity.[27] The wedding took place in a specially modified Boeing 727-200 aircraft, G-Force One, operated by Zero Gravity Corp, or ZERO-G, a company offering weightless flight experiences, of which he is the co-founder.[28]

Other accomplishments and interests

Garriott bought the Luna 21 lander and the Lunokhod 2 rover (both currently on lunar surface) from the Lavochkin Association for $68,500 in December 1993 at a Sotheby's auction in New York[29] (although the catalog incorrectly lists lot 68A as Luna 17/Lunokhod 1).[30]

Garriott acted as corner man for professional boxer and friend Jesús Chávez in his first title defense against Erik Morales in 2004.

He is also an avid magician and magic collector, and appeared on the cover of the January 2008 issue of *MUM*, the magazine of the Society of American Magicians.[31]

Garriott built a haunted house/museum at his residence called Britannia Manor in Austin, Texas. He also promotes private space flight as vice-chairman of the board of directors for Space Adventures and as a trustee of the X-Prize.

Awards

Garriott became the ninth inductee into the Academy of Interactive Arts and Sciences' Hall of Fame in 2006.

Video game works

Game Name	First Released	System Name(s)	Garriott's Role(s)
Akalabeth: World of Doom	1980	Apple II & DOS	Game Designer & Programmer
Ultima I: The First Age of Darkness	1980	Apple II, Atari 8 bit systems, Commodore 64, DOS, FM Towns & MSX	Original Conceptor, Programmer & Graphic Artist
Ultima II: The Revenge of the Enchantress	1982	Apple II, Atari 8 bit systems, Commodore 64, DOS & FM Towns	Programmer
Ultima III: Exodus	1983	Apple II, Amiga, Atari 8 bit systems, Commodore 64, DOS, FM Towns, Mac OS & NES	Project Director
Ultima IV: Quest of the Avatar	1985	Apple II, Amiga, Atari 8 bit systems, Atari ST, Commodore 64, DOS, FM Towns, MSX, NES, Sega Master System	Project Director
Autoduel	1985	Apple II, Amiga, Atari 8 bit systems, Atari ST, Commodore 64 & DOS	Programmer & Designer

Ultima V: Warriors of Destiny	1988	Apple II, Amiga, Atari ST, Commodore 64, DOS, FM Towns & NES	Designer, Writer & Programmer
Omega	1989	Apple II, Amiga, Atari ST, Commodore 64 & DOS	Designer
Ultima VI: The False Prophet	1990	DOS, Amiga, Atari ST, Commodore 64, FM Towns, SNES	Designer, Producer, Sound Effect Worker, Writer & Voice Actor
Worlds of Ultima: The Savage Empire	1990	DOS & SNES	Executive Producer
Ultima: Worlds of Adventure 2: Martian Dreams	1991	DOS	Creative Director
Ultima: Runes of Virtue	1991	Game Boy	Creative Director
Ultima Underworld: The Stygian Abyss	1992	DOS, FM Towns & PlayStation	Director & Voice Actor
Ultima VII: The Black Gate	1992	DOS & SNES	Director & Producer
Ultima VII: The Forge of Virtue	1993	DOS	Creative Assistance & Producer
Ultima VII Part Two: Serpent Isle	1993	DOS	Creative Director & Audio Team Member
Ultima VII Part Two: The Silver Seed	1993	DOS	Director & Voice Actor
Ultima Underworld II: Labyrinth of Worlds	1993	DOS, FM Towns & NEC PC-9801	Director & Voice Actor
Ultima VIII: Pagan	1994	DOS	Producer
Ultima: Runes of Virtue II	1994	Game Boy	Creative Director & Additional Design
Ultima VIII: The Lost Vale Expansion Pack	Cancelled	DOS	Producer
Bioforge	1995	DOS	Executive Producer
Ultima Online	1997	Windows	Producer
Ultima Online: The Second Age	1998	Windows	Executive Designer
Lineage	1998	Windows & Mac OS X	Executive Producer
Ultima IX: Ascension	1999	Windows	Director
Lineage II: The Chaotic Chronicle	2003	Windows	Executive Producer
City of Heroes	2004	Windows	Executive Producer
City of Villains	2005	Windows	Executive Management
Tabula Rasa	Nov. 2 2007	Windows	Executive Producer

External links

- Richard Garriott's Public Web Site [32]
- Blog and details about Garriott's spaceflight [33]
- Spacefacts biography of Richard Garriott [34]
- Richard Garriott's [[Twitter [35]] page]
- Richard Garriott [36] at MobyGames
- Richard Garriott [37] at the Internet Movie Database
- Richard Garriott [38] interview with *Next Generation*
- Richard Garriott [39] interview on UORadio
- Biography of Garriott from DragonCon.org [40]
- Interview with Garriott in Austin Chronicle from 2000 [41]
- Munchkin Village mayor tries to go to space: Computer programmer touts commercial space tourism [42] (The Daily Texan, 7/19/2005)
- Interview with Richard Garriott on space tourism [43] (The Space Review, 8/8/2005)
- The Dot Eaters article [44] featuring a history of Garriott/Lord British, his games, and Origin.
- Interview with Garriott on space and Operation Immortality [45] featuring footage of Richard training.
- Windows on Earth [46]
- MSNBC article entitled "Postcards From Space" [47] about the amateur radio SSTV images sent from the ISS by Richard Garriott
- BBC Breakfast News video interview with Richard Garriott [48] filmed about 10 weeks prior to his space mission.

References

[1] Mark Carreau (2008). "$30 million buys Austin resident a ride on Soyuz mission" (http://www.chron.com/disp/story.mpl/front/6052360.html). The Houston Chronicle. . Retrieved October 12, 2008.
[2] Tariq Malik. "Former Astronaut's Son Signs on as Next Space Tourist" (http://www.space.com/news/070928_garriott_spacetourist.html). SPACE.com. . Retrieved October 9, 2007.
[3] "Inside Ultima IV" (http://cgw.vintagegaming.org/galleries/index.php?year=1986&pub=2&id=26), *Computer Gaming World*: 18–21, March 1986,
[4] King, Brad; John Borland (2003). *Dungeons and Dreamers: The Rise of Computer Game Culture from Geek to Chic*. McGraw-Hill. pp. 11–12. ISBN 0-07-222888-1.
[5] BBC, News (2008-07-15). "BBC NEWS - Science/Nature - Private mission set for ISS" (http://news.bbc.co.uk/1/hi/sci/tech/7507052.stm). . Retrieved 2008-11-11.
[6] Ackerman, Kyle (2002-03-17). "Interview with Richard Garriott at Frictionless Insight" (http://www.frictionlessinsight.com/archives/2002/03/interview-with.html). . Retrieved 2008-01-31.
[7] Safko, Ron; Brake, David (2009). *The Social Media Bible: Tactics, Tools, and Strategies for Business Success*. Wiley. ISBN 0470411554. "Richard Garriott first coined the term *MMORPG* in 1997."
[8] Linder, Brian (2001-05-10). "IGN: Harry Potter LEGO Redux" (http://movies.ign.com/articles/200/200049p1.html). . Retrieved 2007-04-28.
[9] Matonis, Misty (2002-01-05). "When Kings Fall: Part II of II" (http://archive.gamespy.com/mmog/stratics/january02/stratics25/index2.shtm). . Retrieved 2007-04-28.
[10] http://cdn2.libsyn.com/gamepolitics/Garriott-NCsoft-complaint.pdf?nvb=20090509051652&nva=20090510052652&t=049ea4934066248c20f32
[11] Garriott, Richard (2008-11-11). "Richard Garriott's Tabula Rasa - An Open Letter from General British" (http://eu.rgtr.com/en/news_article/an_open_letter_from_general_british). . Retrieved 2008-11-11.
[12] Space Adventures, Ltd. (2008). "Space Adventures Announces 1st Second Generation Astronaut" (http://www.spaceadventures.com/index.cfm?fuseaction=news.viewnews&newsid=554). Space Adventures, Ltd.. . Retrieved October 12, 2008.
[13] Chris Bergin (2008). "Soyuz TMA-13 launches trio on journey to the ISS" (http://www.nasaspaceflight.com/2008/10/soyuz-tma-13-set-to-launch-trio-to-iss/). NASA Spaceflight.com. . Retrieved October 12, 2008.
[14] The Associated Press (2008). "Rocket launches on space station voyage" (http://www.iht.com/articles/ap/2008/10/12/asia/AS-Kazakhstan-Russia-Space.php). International Herald Tribune. . Retrieved October 12, 2008.
[15] Second person to wear the Union flag in space (http://www.timesonline.co.uk/tol/news/world/article4930136.ece)
[16] (http://www.geocaching.com/seek/cache_details.aspx?wp=GC1BE91)

[17] Peter Leonard for The Associated Press. "US game designer blasts into space with DNA cargo, fulfilling childhood dream" (http://www.startribune.com/science/30861229.html). Minneapolis Star Tribune. . Retrieved October 13, 2008.
[18] Marcia Dunn for The Associated Press (2008). "Space tourist will pay high price for adventure" (http://www.msnbc.msn.com/id/25935535). MSNBC. . Retrieved October 12, 2008.
[19] Frank H. Bauer (2008-09-19). "Richard Garriott on ISS" (http://www.southgatearc.org/news/september2008/richard_garriott_on_iss.htm). Southgate Amateur Radio Club. . Retrieved 2008-10-07.
[20] http://ariss-sstv.ssl.berkeley.edu/SSTV/
[21] TERC (2008). "Richard Garriott's Mission in October, 2008" (http://winearth.terc.edu/garriott_mission/). Technical Education Research Centers. . Retrieved October 12, 2008.
[22] "LTUE, Day 2." (http://www.nathanshumate.com/?p=2092). Tachyon City (Nathan Shumate). . Retrieved 2009-02-20.
[23] news.bbc.co.uk, Soyuz space capsule lands safely (http://news.bbc.co.uk/2/hi/science/nature/7688021.stm)
[24] ap.google.com, American, Russians return from space station (http://ap.google.com/article/ALeqM5gZt_AYugND2J2vtj8iqD4H8aQ0fAD940NB780)
[25] en.rian.ru/russia, Soyuz capsule safely lands in Kazakhstan - 2 (http://en.rian.ru/russia/20081024/117925644.html)
[26] itar-tass.com, Cosmonauts examined with special system (http://www.itar-tass.com/eng/level2.html?NewsID=13204951&PageNum=0)
[27] Boyle, Christina (3 June 2009), "So in love they could float away: Brooklyn couple to wed in zero gravity" (http://www.nydailynews.com/ny_local/brooklyn/2009/06/03/2009-06-03_so_in_love_they_could_float_away_bklyn_couple_to_wed_in_zero_gravity.html), *New York Daily News*,
[28] www.reuters.com, Couple floats into zero gravity nuptials (http://www.reuters.com/article/scienceNews/idUSTRE55J22620090621)
[29] The Bloc on the Block (by Jeffrey Kluger): Discover magazine, April 2004 (http://discovermagazine.com/1994/apr/theblocontheblоc363)
[30] Sotheby's Catalogue - *Russian Space History*, Addendum, Lot 68A, December 11, 1993
[31] http://www.cs4fn.org/spacetravel/thespacegame.php
[32] http://www.richardgarriott.com/
[33] http://www.richardinspace.com/
[34] http://www.spacefacts.de/bios/astronauts/english/garriott_richard.htm
[35] http://twitter.com/RichardGarriott
[36] http://www.mobygames.com/developer/sheet/view/developerId,483/
[37] http://www.imdb.com/name/nm0308367/
[38] http://www.next-gen.biz/features/qa-richard-garriotts-tabula-rasa
[39] http://www.box.net/shared/wb3nop2o8c
[40] http://www.dragoncon.org/people/garrior.html
[41] http://www.austinchronicle.com/issues/dispatch/2000-03-03/screens_feature7.html
[42] http://www.dailytexanonline.com/media/paper410/news/2005/07/19/TopStories/Munchkin.Village.Mayor.Tries.To.Go.To.Space-963150.shtml?page=1
[43] http://www.thespacereview.com/article/427/1
[44] http://www.thedoteaters.com/p4_stage2.php
[45] http://www.klru.org/docubloggers/?p=375
[46] http://winearth.terc.edu
[47] http://cosmiclog.msnbc.msn.com/archive/2008/10/23/1577741.aspx
[48] http://news.bbc.co.uk/1/hi/sci/tech/7507052.stm

Guy Laliberté

Guy Laliberté	
 Laliberté at the $25,000 World Poker Tour No Limit Hold'em Championship.	
Born	September 2, 1959 Quebec City, Quebec, Canada
Occupation	Chief Executive Officer
Net worth	▲$2.5 billion 2010[1]
Website Cirque du Soleil [2]	

Guy Laliberté, OC, CQ (born September 2, 1959) is a French Canadian entrepreneur, philanthropist, poker player, space tourist and currently the CEO of Cirque du Soleil. Starting out as an accordion player, stiltwalker and fire-eater; Laliberté created his circus which is a synthesis of all circus styles around the world. In 2006, the 95% share holder of the US $1.2 billion Cirque Du Soleil was named the Ernst & Young Entrepreneur of the Year.[3] [4]

Early years

Laliberté was born in Québec City in 1959. An accordionist, stilt-walker and fire-eater, he founded Quebec's first internationally-renowned circus with the support of a small group of colleagues. A bold visionary, Guy Laliberté recognized and cultivated the talents of the street performers from the Fête foraine de Baie-Saint-Paul and created Cirque du Soleil in 1984.

Guy Laliberté was the first to orchestrate the marriage of cultures and artistic and acrobatic disciplines that is the hallmark of Cirque du Soleil. Since 1984, he has guided the creative team through the creation of every show.

Cirque du Soleil has become a sprawling international operation, as much in terms of its makeup as in the scope of its activities and influence. Guy Laliberté now heads an organization with activities on five continents. [5]

Personal life

Guy Laliberté	

Spaceflight Participant	
Born	2 September 1959 Quebec City, Quebec
Other occupation	Entrepreneur
Time in space	10d 21h 17m
Missions	1 mission consisting of Soyuz TMA-16 (up) & Soyuz TMA-14 (return)
Mission insignia	

Relationships

Guy Laliberté has been romantically linked with Montreal fashion model Claudia Barilla with whom he has two children. He had previously fathered three children with Rizia Moreira, a Brazilian woman 15 years younger than he.

Spaceflight experience

In September 2009, Guy Laliberté became the first Canadian private space explorer. His mission was dedicated to raising awareness on water issues facing humankind on planet Earth. Under the theme Moving Stars and Earth for Water, this first Poetic Social Mission in space aimed at touching people through an artistic approach: a special 120-minute webcast program featuring various artistic performances unfolding in 14 cities on five continents, including the International Space Station. [5]

Poker career

In April 2007, Laliberté finished fourth in the World Poker Tour Season Five event at Bellagio in Las Vegas and won $696,220. Laliberté also played on GSN's High Stakes Poker Season 4 show. He also took part in Poker After Dark season 4.

Philanthropy

October 29, 2007, Guy Laliberté announced, in the presence of H.S.H. Prince Albert II of Monaco, Jeremy Hobbs, Executive Director of Oxfam International and Gordon Nixon, President and Chief Executive Officer of RBC Financial Group, the official launch of the ONE DROP Foundation[6] to fight poverty in the world by giving everyone access to water. Inspired by the creative experience of Cirque du Soleil and its international program for street kids, Cirque du Monde, the ONE DROP Foundation makes use of the circus arts, folklore, popular theatre, music, dance and the visual arts to promote education, community involvement and public awareness of water issues. Technical projects in developing countries will improve access to water, ensure food security and promote gender equality in communities.

The operating costs of ONE DROP will be covered by a $100 million contribution from Guy Laliberté over 25 years. Field activities will be financed by donations from the employees of Cirque du Soleil and from the public, as well as through funding commitments by Canadian and international partners. Oxfam International, through Oxfam-Québec, has been associated with ONE DROP since 2005 in a three-year pilot project in Nicaragua. A leader in development aid, Oxfam brings expertise in selecting and implementing field projects. Their involvement with the ONE DROP Foundation is based on a common desire to support sustainable development with concrete actions and in collaboration with local partners.

In March 2009, Guy Laliberté was rumoured as one of the potential buyers of the Montreal Canadiens hockey club, as former owner George Gillett explored the option of selling the franchise.[7]

Awards and honours

Université Laval (Québec) awarded an honorary Doctorate to Guy Laliberté in 2008. The year before, Guy Laliberté took the Ernst & Young Entrepreneur of the Year award for all three levels: Quebec, Canada and international. In 2004, he received the Order of Canada, the highest distinction in the country, from the Governor General of Canada. The same year, he was recognized by Time Magazine as one of the 100 most influential people in the world. In 2003, he was honoured by the Condé Nast group as part of the Never Follow Program, a tribute to creators and innovators. In 2001, he was named a Great Montrealer by the Académie des Grands Montréalais. In 1997, Guy Laliberté received the Ordre National du Québec, the highest distinction awarded by the Government of Quebec. [5]

External links

- Cirque du Soleil's Official Website [8]
- One Drop Foundation [9]
- Spacefacts biography of Guy Laliberté [10]
- Forbes topic page on Guy Laliberte [11]

References

[1] Forbes topic page on Guy Laliberte (http://billionaires.forbes.com/topic/guy_laliberte) Forbes.com. Accessed May 2010.
[2] http://www.cirquedusoleil.com/
[3] Ernst & Young, 2006 EOY National Recipient (http://www.ey.com/global/content.nsf/Canada/Media_-_2006_-_EOY_National_Recipient)
[4] Time Magazine, Guy Laliberte (http://www.time.com/time/magazine/article/0,9171,994013,00.html)
[5] http://www.cirquedusoleil.com/CirqueDuSoleil/en/Pressroom/cirquedusoleil/biographies/Laliberte_Guy.htm
[6] Onedrop.org (http://www.onedrop.org/en/journal/detail/624.htm)
[7] Stubbs, Dave (2009-03-23). "Players not worried about possible Canadiens' sale" (http://www.faceoff.com/story.html?id=a04cce42-f515-4e06-b99f-d8ed9ee21ca3). Faceoff. . Retrieved 2009-04-02.
[8] http://www.cirquedusoleil.com
[9] http://www.onedrop.org/en
[10] http://www.spacefacts.de/bios/international/english/laliberte_guy.htm
[11] http://billionaires.forbes.com/topic/guy_laliberte

Private spaceflight

Astronaut Dale A. Gardner holding a "For Sale" sign

Private spaceflight is flight above 100 km (62 mi) Earth altitude conducted by and paid for by an entity other than a government. In the early decades of the Space Age, the government space agencies of the Soviet Union and United States pioneered space technology augmented by collaboration with affiliated design bureaus in the USSR and private companies in the US. The European Space Agency was formed in 1975, largely following the same model of space technology development. Later on, large defense contractors began to develop and operate space launch systems, derived from government rockets and commercial satellites. Private spaceflight in Earth orbit includes communications satellites, satellite television, satellite radio, astronaut transport and sub-orbital and orbital space tourism. Recently, entrepreneurs have begun designing and deploying competitive space systems to the national-monopoly governmental systems of the early decades of the space age. Successes to date include flying suborbital spaceplanes and launching lightweight orbital rockets. Planned private spaceflights beyond Earth orbit include solar sailing prototypes, deep space burial and personal spaceflights around the Moon. A private orbital habitat prototype is already in Earth orbit, with larger versions to follow.[1]

History of commercial space transportation

Launch of a Proton rocket

During the early years of spaceflight only nation states had the resources to develop and fly spacecraft. Both the U.S. space program and Soviet space program were operated using mainly military pilots as astronauts. During this period, no commercial space launches were available to private operators, and no private organization was able to offer space launches. Eventually, private organizations were able to both offer and purchase space launches, thus beginning the period of private spaceflight.

The first phase of private space operation was the launch of the first commercial communications satellites. The U.S. Communications Satellite Act of 1962 opened the way to commercial consortia owning and operating their own satellites, although these were still launched on state-owned launch vehicles.

European sponsorship

On March 26, 1980, the European Space Agency created Arianespace, the world's first commercial space transportation company. Arianespace produces, operates and markets the Ariane launcher family. By 1995 Arianespace lofted its 100th satellite and by 1997 the Ariane rocket had its 100th launch.[2] Arianespace's 23 shareholders represent scientific, technical, financial and political entities from 10 different European countries.[3]

American deregulation

From the beginning of the Shuttle program until the Challenger disaster in 1986, it was the policy of the United States that NASA be the public-sector provider of U.S. launch capacity to the world market.[4] Initially NASA subsidized satellite launches with the intention of eventually pricing Shuttle service for the commercial market at long-run marginal cost.

On October 30, 1984, United States President Ronald Reagan signed into law the Commercial Space Launch Act.[5] This enabled an American industry of private operators of expendable launch systems. Prior to the signing of this law, all commercial satellite launches in the United States were restricted by Federal regulation to NASA's Space Shuttle.

On November 5, 1990, United States President George H. W. Bush signed into law the Launch Services Purchase Act.[6] The Act, in a complete reversal of the earlier Space Shuttle monopoly, ordered NASA to purchase launch services for its primary payloads from commercial providers whenever such services are required in the course of its activities.

Commercial launches outnumbered government launches at the Eastern Test Range in 1997.[7]

Russian privatization

The Russian government sold part of its stake in RSC Energia to private investors in 1994. Energia together with Khrunichev constituted most of the Russian manned space program. In 1997, the Russian government sold off enough of its share to lose the majority position.

American subsidization

In 1996 the United States government selected Lockheed Martin and Boeing to each develop Evolved Expendable Launch Vehicles (EELV) to compete for launch contracts and provide assured access to space. The government's acquisition strategy relied on the strong commercial viability of both vehicles to lower unit costs. This anticipated market demand did not materialize, but both the Delta IV and Atlas V EELVs remain in active service.

Launch alliances

Since 1995 Khrunichev's Proton rocket is marketed through International Launch Services while the Soyuz rocket is marketed via Starsem. Energia builds the Soyuz rocket and owns part of the Sea Launch project which flies the Ukrainian Zenit rocket.

In 2003 Arianespace joined with Boeing Launch Services and Mitsubishi Heavy Industries to create the Launch Services Alliance. In 2005, continued weak commercial demand for EELV launches drove Lockheed Martin and Boeing to propose a joint venture called the United Launch Alliance to service the United States government launch market.[8]

Human spaceflight privatization

On February 1, 2010 United States President Barack Obama proposed in a speech that NASA exit the business of flying astronauts from Earth to orbit. The proposal acted on the findings of the 2009 Augustine Commission and built on the success of the Commercial Resupply Services that outsourced American cargo delivery to the International Space Station.[9]

Today many commercial space transportation companies offer launch services to satellite companies and government space organizations around the world. In 2005 there were 18 total commercial launches and 37 non-commercial launches.[10] Russia flew 44% of commercial orbital launches, while Europe had 28% and the United States had 6%.

Private spaceflight companies

Commercial launchers

The space transport business serves primarily national government and large commercial customer segments. Launches of government payloads, including military, civilian and scientific satellites, is the largest market segment at nearly $100 billion a year. This segment is dominated by domestic favorites such as the United Launch Alliance for U.S. government payloads and Arianespace for European satellites. The commercial payload segment, valued at under $3 billion a year, is dominated by Arianespace, with over 50% of the market segment,[11] followed by Russian launchers. See a complete list of launch systems.

Commercial Orbital Transportation Services

On January 18, 2006 NASA announced an opportunity for commercial providers to demonstrate orbital transportation services.[12] NASA plans to spend $500 million through 2010 to finance development of private sector capability to transport payloads to the International Space Station (ISS). This is more challenging than extant commercial space transportation because it requires precision orbit insertion, rendezvous and possibly docking with another spacecraft. The commercial vendors will compete in specific service areas.[13] NASA Administrator Michael D. Griffin has stated that without affordable commercial orbital transportation services (COTS), the agency will not have enough funds remaining to achieve the objectives of the Vision for Space Exploration.

Computer rendering of SpaceX Dragon docking with the ISS

In August 2006, NASA announced that two fledgling aerospace companies, SpaceX and Rocketplane Kistler, had been awarded $278m and $207m, respectively, under the COTS program.[14] NASA anticipates that COTS services to ISS will be necessary through at least 2015. The NASA Administrator has suggested that space transportation services procurement may be expanded to orbital fuel depots and lunar surface deliveries should the first phase of COTS prove successful.[15]

After it transpired that Rocketplane Kistler was failing to meet its deadlines, the NASA terminated their contract in August 2008, after only $32m had been spent. Several months later, in December 2008, NASA announced that they have awarded the remaining $170m to the trusted Orbital Sciences Corporation to develop resupply services to the ISS.[16]

Emerging personal spaceflight

SpaceShipOne has a 5-meter wingspan and a 3-person cabin.

Before 2004 no privately operated manned spaceflight had ever occurred. The only private individuals to journey to space went as space tourists in the Space Shuttle or on Russian Soyuz flights to Mir or the International Space Station.

All private individuals who flew to space before Dennis Tito's self-financed International Space Station visit in 2001 had been sponsored by their home governments. Those trips include US Congressman Bill Nelson's January 1986 flight on the Space Shuttle Columbia and Japanese television reporter Toyohiro Akiyama's 1990 flight to the Mir Space Station.

The Ansari X PRIZE was intended to stimulate private investment in the development of spaceflight technologies. The June 21, 2004 test flight of SpaceShipOne, a contender for the X PRIZE, was the first human spaceflight in a privately developed and operated vehicle.

On September 27, 2004, following the success of SpaceShipOne, Richard Branson, owner of Virgin and Burt Rutan, SpaceShipOne's designer, announced that Virgin Galactic had licensed the craft's technology, and were planning commercial space flights in 2.5 to 3 years. A fleet of five craft (SpaceShipTwo, launched from the WhiteKnightTwo carrier airplane) is to be constructed, and flights will be offered at around $200,000 each, although Branson has said he plans to use this money to make flights more affordable in the long term.

XCOR Aerospace also plans to initiate a suborbital commercial spaceflight service with the Lynx rocketplane in 2012. First test flights are planned for 2010.

In December 2004, United States President George W. Bush signed in to law the Commercial Space Launch Amendments Act.[17] The Act resolved the regulatory ambiguity surrounding private spaceflights and is designed to promote the development of the emerging U.S. commercial human space flight industry.

On July 12, 2006, Bigelow Aerospace launched the *Genesis I*, a subscale pathfinder of an orbital space station module. *Genesis II* was launched on June 28, 2007, and there are plans for additional prototypes to be launched in preparation for the production model *BA 330* spacecraft.

On September 28, 2006, Jim Benson, SpaceDev founder, announced he was founding Benson Space Company with the intention of being first to market with the safest and lowest cost suborbital personal spaceflight launches, using the vertical takeoff and horizontal landing Dream Chaser vehicle based on the NASA HL-20 Personnel Launch System vehicle.

Failed spaceflight ventures

In the 1990s the projection of a significant demand for communications satellite launches attracted the development of a number of commercial space launch providers. The launch demand largely vanished when some of the largest satellite constellations, such as 288 satellite Teledesic network, were never built. The historic tendency of NASA to compete against the private sector and the Department of Defense's preference for the traditional military industrial complex has discouraged many new space launch ventures.

VentureStar

Lockheed Martin VentureStar

In 1996 NASA selected Lockheed Martin Skunk Works to build the X-33 VentureStar prototype for a single stage to orbit (SSTO) reusable launch vehicle. In 1999, the subscale X-33 prototype's composite liquid hydrogen fuel tank failed during testing. At project termination on March 31, 2001, NASA had funded $912 million of this wedge shaped spacecraft while Lockheed Martin financed $357 million of it.[18] The VentureStar was to have been a full-scale commercial space transport operated by Lockheed Martin.

Beal Aerospace

In 1997 Beal Aerospace proposed the BA-2, a low-cost heavy-lift commercial launch vehicle. In March 4, 2000, the BA-2 project tested the largest liquid rocket engine built since the Saturn V.[19] In October 2000, Beal Aerospace ceased operations citing a decision by NASA and the Department of Defense to commit themselves to the development of the competing government-financed EELV program.[20]

Rotary Rocket

In 1998 Rotary Rocket proposed the Roton, a Single Stage to Orbit (SSTO) piloted Vertical Take-off and Landing (VTOL) space transport.[21] A full scale Roton Atmospheric Test Vehicle flew three times in 1999. After spending tens of millions of dollars in development the Roton failed to secure launch contracts and Rotary Rocket ceased operations in 2001.

Future plans

Many have speculated on where private spaceflight may go in the near future. One possibility is for paid suborbital tourism on craft like SpaceShipOne. Additionally, suborbital spacecraft have applications for faster intercontinental package delivery and passenger flight.

Private orbital spaceflight, space stations

SpaceX's Falcon 9 rocket, scheduled to be first launched in 2009, is designed to be man-rated. This would be the first American orbital vehicle since the Space Shuttle to receive this designation, in principle allowing the vehicle to transport paying customers to orbit. Plans and a full-scale prototype for the SpaceX Dragon, a manned capsule carrying up to 7 passengers, were announced on March 6, 2006.[22]

An early flight of the Falcon 9 is planned to carry Sundancer, the prototype expandable and habitable space module (based on the former NASA TransHab design) constructed by Bigelow Aerospace. Bigelow Aerospace expects modules like Sundancer and the larger BA 330 to be used for activities like microgravity research, space manufacturing, and space tourism (with modules serving as orbital hotels). To promote private manned launch efforts, Bigelow offered the US$50 million America's Space Prize for the first US-based privately funded team to launch a manned reusable spacecraft to orbit on or before January 10, 2010, though no company was able to meet this deadline.

Excalibur Almaz plans to launch a modernized Almaz space station, for tourism and other uses. It will feature the largest window ever on a spacecraft.

On-orbit propellant depots

In a presentation given November 15, 2005, to the 52nd Annual Conference of the American Astronautical Society, NASA Administrator Michael D. Griffin suggested that establishing an on-orbit propellant depot is, "Exactly the type of enterprise which should be left to industry and to the marketplace."[23] At the Space Technology and Applications International Forum in 2007, Dallas Bienhoff of Boeing made a presentation detailing the benefits of propellant depots.[24]

Asteroid mining

Some have speculated on the profitability of mining metal from asteroids. According to some estimates, a one kilometer-diameter asteroid would contain 30 million tons of nickel, 1.5 million tons of metal cobalt and 7,500 tons of platinum; the platinum alone would have a value of more than $500 billion at current prices.[25] While the potential rewards from asteroid mining are indeed huge, the technical challenges are equally large and it seems likely that the private sector will wait for the publicly funded space program to solve them (e.g. by establishing experimental mines on the Moon).

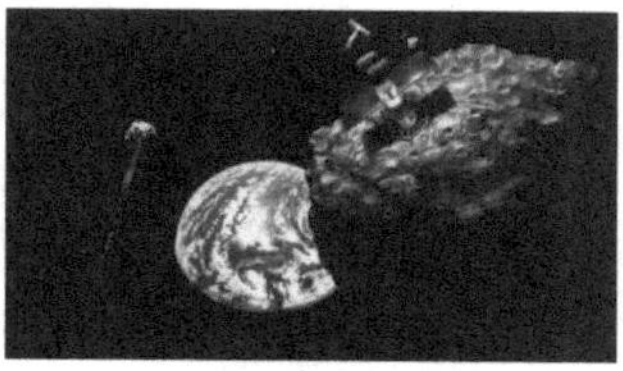

Asteroid mining spacecraft

Energy from space

Future energy development may use energy sources in space and on other planets. Examples include Helium-3 extraction from the Moon, and solar power satellite systems. See space manufacturing for more on extraterrestrial economic development.

Space elevators

A Space Elevator system is a possible launch system, currently under investigation by at least one private venture.[26] There are concerns over cost, general feasibility and some political issues. On the plus side the potential to scale the system to accommodate traffic would (in theory) be greater than some other alternatives. Some factions contend that a space elevator — if successful — would not supplant existing launch solutions but complement them.

See also

- Commercial Spaceflight Federation
- Space Frontier Foundation
- L5 Society
- Blue Origin
- Heinlein Prize for Advances in Space Commercialization
- Masten Space Systems
- X Prize Foundation
- XCOR Aerospace

References

[1] "Special Announcement" (http://www.bigelowaerospace.com/news/?Special_Announcement). bigelowaerospace.com. . Retrieved 2008-04-01.

[2] "Milestones" (http://web.archive.org/web/20080113040347/http://www.arianespace.com/site/about/milestones_sub_index.html). *Arianespace.com*. Archived from the original (http://www.arianespace.com/site/about/milestones_sub_index.html) on 2008-01-13. . Retrieved 2008-02-14.

[3] "Arianespace shareholders represent scientific, technical, financial and political entities from 10 different European countries" (http://web.archive.org/web/20080206075437/http://www.arianespace.com/site/about/shareholders_sub_index.html). *Arianespace.com*. Archived from the original (http://www.arianespace.com/site/about/shareholders_sub_index.html) on 2008-02-06. . Retrieved 2008-02-14.

[4] "Setting Space Transportation Policy for the 1990s" (http://www.cbo.gov/ftpdocs/59xx/doc5935/doc24c-Entire.pdf) (PDF). cbo.gov. . Retrieved 2008-02-14.

[5] "Statement on Signing the Commercial Space Launch Act" (http://www.reagan.utexas.edu/archives/speeches/1984/103084i.htm). reagan.utexas.edu. . Retrieved 2008-02-14.

[6] "$ 2465d. Requirement of US Federal government to procure commercial launch services" (http://www.space-frontier.org/commercialspace/lspalaw.txt). space-frontier.org. . Retrieved 2008-02-14.

[7] "Streamlining Space Launch Range Safety - Executive Summary" (http://books.nap.edu/openbook.php?record_id=9790&page=1). National Academy of Sciences. . Retrieved 2008-02-13.

[8] "Boeing, Lockheed Martin to Form Launch Services Joint Venture" (http://www.spaceref.com/news/viewpr.html?pid=16790). spaceref.com. . Retrieved 2008-02-13.

[9] Chang, Kenneth. "Obama Calls for End to NASA's Moon Program" (http://www.nytimes.com/2010/02/02/science/space/02nasa.html) (PDF). nytimes.com. . Retrieved 2010-02-01.

[10] "Commercial Space Transportation: 2005 Year In Review" (http://www.faa.gov/about/office_org/headquarters_offices/ast/media/2005_YIR_FAA_AST_0206.pdf) (PDF). faa.gov. . Retrieved 2008-02-13.

[11] "Changing Trajectory: French Firms Vaults Ahead in Civilian Rocket Market". *The Wall Street Journal* (Dow Jones & Company, Inc.): A1. June 25, 2007.

[12] "NASA Seeks Proposals for Crew and Cargo Transportation to Orbit" (http://www.spaceref.com/news/viewpr.html?pid=18791). spaceref.com. . Retrieved 2008-02-11.

[13] "GUIDANCE_FOR_THE_PREPARATION_AND_SUBMISSION_OF_UNSOLICITED_PROPOSALS" (http://ec.msfc.nasa.gov/hq/library/unSol-Prop.html). nasa.gov. . Retrieved 2008-02-13.

[14] "NASA Invests in Private Sector Space Flight with SpaceX, Rocketplane-Kistler" (http://www.nasa.gov/mission_pages/exploration/news/COTS_selection.html). nasa.gov. . Retrieved 2008-02-14.

[15] "Human_Space_Flight_Transition_Plan" (http://spaceoperations.nasa.gov/tran_plan.pdf) (PDF). nasa.gov. . Retrieved 2008-02-14.

[16] "NASA Awards Space Station Commercial Resupply Services Contracts" (http://www.nasa.gov/home/hqnews/2008/dec/HQ_C08-069_ISS_Resupply.html). nasa.gov. . Retrieved 2008-12-24.

[17] "House Approves H.R. 3752, The Commercial Space Launch Amendments Act of 2004" (http://www.spaceref.com/news/viewpr.html?pid=13774). spaceref.com. . Retrieved 2008-02-14.

[18] "NASA Reaches Milestone in Space Launch Initiative Program" (http://www.hq.nasa.gov/office/pao/History/x-33/01-31.htm). nasa.gov. . Retrieved 2008-02-14.

[19] "Beal BA-2" (http://www.astronautix.com/lvs/bealba2.htm). astronautix.com. . Retrieved 2008-02-14.

[20] spaceprojects.com (2000-03-23). "Beal Aerospace regrets to announce that it is ceasing all business operations effective October 23, 2000" (http://www.spaceprojects.com/Beal/). Press release. . Retrieved 2007-06-21.

[21] "Roton" (http://www.astronautix.com/craft/roton.htm). astronautix.com. . Retrieved 2008-02-14.

[22] Cowing, Keith (2006-03-06). "The SpaceX Dragon: America's First Privately Financed Manned Orbital Spacecraft?" (http://www.spaceref.com/news/viewnews.html?id=1095). SpaceRef.com. . Retrieved 2008-02-12.

[23] "NASA and the Business of Space" (http://www.nasa.gov/pdf/138033main_griffin_aas1.pdf) (PDF). NASA. . Retrieved 2008-02-12.

[24] "The Potential Impact of a LEO Propellant Depot on the NASA ESAS Architecture" (http://www.boeing.com/defense-space/space/constellation/references/presentations/Potential_Impact_of_LEO_Propellant_on_NASA_ESAS_Architecture.pdf) (PDF). Boeing. . Retrieved 2008-02-12.

[25] "How Asteroid Mining Will Work" (http://science.howstuffworks.com/asteroid-mining1.htm). howstuffworks.com. . Retrieved 2008-02-12.

[26] "The LiftPort Space Elevator" (http://www.liftport.com/). liftport.com. . Retrieved 2008-02-11.

- Belfiore, Michael. *Rocketeers: How a Visionary Band of Business Leaders, Engineers, and Pilots is Boldly Privatizing Space*. Harper Paperbacks, 2008.
- Bizony, Piers. *How to Build Your Own Spaceship: The Science of Personal Space Travel*. Plume, 2009.

External links

- Climbing a Commercial Stairway to Space: A Plausible Timeline (http://www.hobbyspace.com/AAdmin/archive/SpecialTopics/toSpaceTimeLine.html) RLV News, February 2, 2006
- An Introduction to Private Spaceflight (http://spaceliberates.us/news/an-introduction-to-private-spaceflight/) Space Liberates Us!, March 20, 2007
- Study defining personal spaceflight industry (http://spacefellowship.com/News/?p=5480) Space Fellowship, May 29, 2008

Government

- Suborbital Reusable Launch Vehicles and Applicable Markets (http://www.nesdis.noaa.gov/space/library/reports/2002-10-suborbital-LowRes.pdf) DOC OSC, 2002

Corporate ventures

- C&SPACE (http://www.candspace.com/)
- Starchaser Industries (http://www.starchaser.co.uk/)
- Space Services, Inc. (http://www.spaceservicesinc.com/)
- Commercial Space Companies (http://www.space-frontier.org/commercialspace/) at the Space Frontier Foundation

Media coverage

- A Word from the Know-Nothing Bureaucrats (http://www.spaceref.com/news/viewsr.html?pid=12252) NASA union viewpoint on private spaceflight
- Private Industry Can Help NASA Open the Space Frontier (http://www.space-frontier.org/OpEds/20050214privateindustry.html) Space Frontier Foundation, February 14, 2005

SpaceShipOne

SpaceShipOne	

Astronaut Mike Melvill after his September 29, 2004 spaceflight.	
Role	Spaceplane
Manufacturer	Scaled Composites
Designed by	Burt Rutan
First flight	20 May 2003
Retired	4 October 2004
Status	Museum piece
Primary user	Mojave Aerospace Ventures
Number built	1
Variants	SpaceShipTwo

SpaceShipOne is a spaceplane that completed the first privately funded spaceflight on June 21, 2004.

(L to R) Marion Blakely (FAA), Mike Melvill; Richard Branson, Burt Rutan, Brian Binnie, and Paul Allen reflect on a misson accomplished (October 4, 2004)

SpaceShipOne was developed by Mojave Aerospace Ventures (a joint venture between Paul Allen and Scaled Composites, Burt Rutan's aviation company, in their Tier One program), without government funding. On June 21, 2004, it made the first privately funded human spaceflight. On October 4, it won the $10 million Ansari X PRIZE, by reaching 100 kilometers in altitude twice in a two-week period with the equivalent of three people on board and with no more than ten percent of the non-fuel weight of the spacecraft replaced between flights. Development costs were estimated to be $25 million, funded completely by Paul Allen.

During its test programme, SpaceShipOne set a number of important "firsts", including first privately funded aircraft to exceed Mach 2 and Mach 3, first privately funded manned spacecraft to exceed 100km altitude, and first privately funded reusable manned spacecraft.

SpaceShipOne is an experimental air-launched rocket-powered aircraft with suborbital flight capability that uses a hybrid rocket motor. The design features a unique "feathering" atmospheric reentry system where the rear half of the wing and the twin tail booms folded upward along a hinge running the length of the wing; this increased drag while remaining stable. The achievements of SpaceShipOne are more comparable to the X-15 than orbiting spacecraft like the Space Shuttle. Accelerating a spacecraft to orbital speed requires more than 60 times as much energy as accelerating it to Mach 3.[1]

History

SpaceShipOne is registered with the FAA as **N328KF** [2]. *N* is the prefix for US-registered aircraft; *328KF* was chosen by Scaled Composites to stand for 328 kilo feet (about 100 kilometers), the officially designated edge of space. The original choice of registry number, **N100KM**, was already taken. N328KF is registered as a glider, reflecting the fact that most of its independent flight is unpowered.

SpaceShipOne landing after its June 21, 2004 space flight (Flight 15P)

SpaceShipOne's first flight, 01C, was an unmanned captive carry flight test on May 20, 2003. Glide tests followed, starting with flight 03G on August 7, 2003. Its first powered flight, flight 11P, was made on December 17, 2003, the 100th anniversary of the first powered flight.

On April 1, 2004, Scaled Composites received the first license for sub-orbital rocket flights to be issued by the US Office of Commercial Space Transportation. This license permitted the company to conduct powered test flights over the course of one year. On June 17, 2004, Mojave Airport reclassified itself (part-time) as the Mojave Spaceport.

Flight 15P on June 21, 2004, was SpaceShipOne's first spaceflight, and the first privately funded human spaceflight. Ansari X PRIZE flights followed, with flight 17P on October 4, 2004, winning the prize.

The SpaceShipOne Team was awarded the Space Achievement Award[3] by the Space Foundation in 2005.

Retirement

SpaceShipOne's spaceflights were watched by large crowds at Mojave Spaceport. A fourth suborbital flight, Flight 18P, was originally scheduled for October 13, 2004. However, Burt Rutan decided not to risk damage to the historic craft, and cancelled it and all future flights.

SpaceShipOne now hangs in the National Air and Space Museum in Washington D.C.

On July 25, 2005 SpaceShipOne was taken to the Oshkosh Airshow in Oshkosh, Wisconsin. After the airshow, Mike Melvill and crew flew the White Knight, carrying SpaceShipOne, to Wright-Patterson Air Force Base in Dayton, Ohio, where Mike spoke to a group of about 300 military and civilian personnel. Later in the evening, Mike gave a presentation at the Dayton Engineers Club, entitled "Some Experiments in Space Flight", in honor of Wilbur Wright's now famous presentation to the American Society of Mechanical Engineers in 1901 entitled "Some Experiments in Flight." The White Knight then transported SpaceShipOne to the Smithsonian Institution's National Air and Space Museum to be put on display. It was unveiled on Wednesday October 5, 2005 in the Milestones of Flight gallery and is now on display to the public in the main atrium between the Spirit of St. Louis and the Bell X-1.

SpaceShipOne became a popular model rocket in 2004. Estes Industries currently offers several flying model rockets of SpaceShipOne. A piece of SpaceShipOne's carbon fiber material was launched aboard the New Horizons mission to Pluto in 2006.[4]

Replica

SpaceShipOne Replica in normal configuration

A year after its appearance in the Oshkosh Airventure airshow, the Experimental Aircraft Association featured a full-scale replica of the spacecraft in a wing of its museum which housed other creations of Burt Rutan. Using the same fiberglass molds as the original, it was so exact in its replication - despite not having any doors or interior - that it was dubbed "Serial 2 Scaled" by Scaled Composites. Each and every painstaking detail in its appearance was matched, down to the N328KF registration number on its fuselage. It is so precise that, during a video presentation held every hour in the museum, it can display the two different modes of its 'feathering' ability, albeit through the aid of pulleys and wires (there is no machinery in the replica).[5]

Another full-scale replica of SpaceShipOne hangs in the rotunda of the William Thomas Terminal at Meadows Field Airport in Bakersfield and a third is on display in the Mojave Spaceport's Legacy Park alongside the original Roton Atmospheric Test Vehicle.

Future efforts

The development of a larger, more powerful space plane capable of reaching 120 km was announced in 2004. It is named Scaled Composites SpaceShipTwo, and is currently under development by The Spaceship Company, a joint venture between Scaled Composites and Sir Richard Branson's Virgin Group, as part of the Tier 1b program. The Virgin Galactic spaceliner plans to operate a fleet of five of these craft in passenger-carrying private spaceflight service starting no earlier than 2011.[6] [7]

WhiteKnightTwo, the carrier aircraft for SpaceShipTwo, was unveiled to the public at the Experimental Aircraft Association's Air Venture in Oshkosh, Wisconsin on July 28, 2009. Following a series of 50 – 100 SpaceShipTwo test flights, the first paying customers are expected to fly aboard the craft in 2011. In August 2005, Virgin Galactic stated that if the upcoming suborbital service with SpaceShipTwo is successful, the follow-up SpaceShipThree will be an orbital craft.[8] [9]

Flights

SpaceShipOne in flight.

All of the flights of SpaceShipOne were from the Mojave Airport Civilian Flight Test Center. Flights were numbered, starting with flight 01 on May 20, 2003. One or two letters are appended to the number to indicate the type of mission. An appended **C** indicates that the flight was a captive carry, **G** indicates an unpowered glide, and **P** indicates a powered flight. If the actual flight differs in category from the intended flight, two letters are appended: the first indicating the intended mission and the second the mission actually performed.

In the table below, the "top speed" reported is the Mach number at burn-out (the end of the rocket burn). This is not an absolute speed.

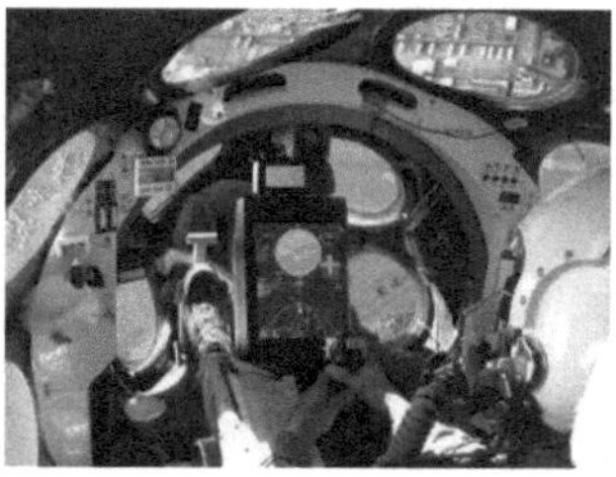

Cockpit of WhiteKnight in flight, EFIS display edited to look like SpaceShipOne.

SpaceShipOne flights

Flight	Date	Top speed	Altitude	Duration	Pilot
01C	May 20, 2003			1 h 48 min	unmanned
02C	July 29, 2003			2 h 06 min	Mike Melvill
03G	August 7, 2003			19 min 00 s	Mike Melvill
04GC	August 27, 2003			1 h 06 min	Mike Melvill
05G	August 27, 2003			10 min 30 s	Mike Melvill
06G	September 23, 2003			12 min 15 s	Mike Melvill
07G	October 17, 2003			17 min 49 s	Mike Melvill
08G	November 14, 2003			19 min 55 s	Peter Siebold
09G	November 19, 2003			12 min 25 s	Mike Melvill
10G	December 4, 2003			13 min 14 s	Brian Binnie
11P	December 17, 2003	Mach 1.2	20.7 km	18 min 10 s	Brian Binnie
12G	March 11, 2004			18 min 30 s	Peter Siebold
13P	April 8, 2004	Mach 1.6	32.0 km	16 min 27 s	Peter Siebold
14P	May 13, 2004	Mach 2.5	64.3 km	20 min 44 s	Mike Melvill
15P	June 21, 2004	Mach 2.9	100.1 km	24 min 05 s	Mike Melvill
16P	September 29, 2004	Mach 2.92	102.9 km	24 min 11 s	Mike Melvill
17P	October 4, 2004	Mach 3.09	112.0 km	23 min 56 s	Brian Binnie

The flights were accompanied by two chase planes; an Extra 300 owned and flown by Chuck Coleman, and a Beechcraft Starship[10] .

Astronauts

The SpaceShipOne pilots came from a variety of aerospace backgrounds. Mike Melvill is a test pilot, Brian Binnie is a former Navy pilot, and Doug Shane and Peter Siebold are engineers at Scaled Composites. They qualified to fly SpaceShipOne by training on the Tier One flight simulator and in White Knight and other Scaled Composites aircraft.

Specifications

Data from astronautix.com[11]

Rear view

General characteristics

- **Crew:** one, pilot
- **Capacity:** 2 passengers
- **Length:** 16 ft 5 in (8,05 m)
- **Wingspan:** 16 ft 5 in (8,05 m)
- **Height:** ()
- **Wing area:** 161.4 ft^2 (15 m^2)

- **Empty weight:** 2,640 lb (1,200 kg)
- **Loaded weight:** 7,920 lb (3,600 kg)
- **Powerplant:** 1× N2O/HTPB SpaceDev Hybrid rocket motor, 7,500 kgf (74 kN)
- **I_{sp}:** 250 s (2450 Ns/kg)
- **Burn time:** 87 seconds
- **Aspect Ratio:** 1.6

Performance

- **Maximum speed:** Mach 3.09 (2,170 mph, 3,518 km/h)
- **Range:** 35 nm (40 mi, 65 km)
- **Service ceiling:** 367,360 ft (112,000 m)
- **Rate of climb:** 82,000 ft/min (416.6 m/s)
- **Wing loading:** 49.07 lb/ft^2 (240 kg/m^2)
- **Thrust/weight:** 2.08

Gallery

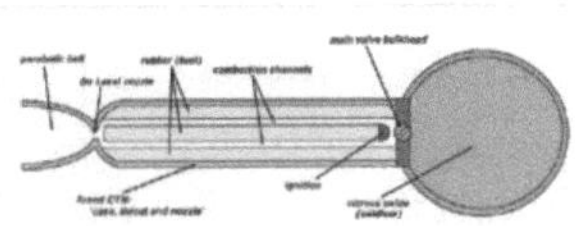

Hybrid rocket engine detail of SpaceShipOne (more information [12]).

SpaceShipOne in the National Air and Space Museum in Washington D.C. with the Spirit of Saint Louis and Bell X-1 "Glamorous Glennis"

See also

- List of spaceflights (2004)
- Mojave Magic: A Turtle's Eye View of SpaceShipOne
- Space Ship One (album)
- X-15

External links

- Tier One home page [13] at Scaled Composites's website
  - Flight profile in PDF format [14] and in JPG format [15]
- SpaceDev [16]
- SpaceShipOne video footage [17] (RealVideo) of flight 14P
- Spaceship One music video [18]
- Wired News: SpaceShipOne Back on Course [19]
- SpaceShipOne's FAA registration [20]
- The Ansari X PRIZE [47]
- SpaceShipOne documentary DVD - *Mojave Magic* [21]
  - Song *I Want To Go In A Space Ship* about SpaceShipOne from *Mojave Magic* [22]
  - Video of Burt Rutan singing the song at the 2005 Oshkosh airshow [23]
- Patent No. 7,195,207 on SpaceShipOne [24]

References

[1] Elon Musk Is Betting His Fortune on a Mission Beyond Earth's Orbit (http://www.wired.com/science/space/magazine/15-06/ff_space_musk?currentPage=3)
[2] http://registry.faa.gov/aircraftinquiry/NNum_Results.aspx?NNumbertxt=328KF
[3] http://www.nationalspacesymposium.org/symposium-awards
[4] "Pluto Mission to Carry Piece of SpaceShipOne" (http://www.space.com/astronotes/astronotes.html). Space.com. December 20, 2005. .
[5] Jeff Hecht (29 July 2005). "SpaceShipOne" (http://www.airventuremuseum.org/collection/aircraft/SpaceShipOne.asp). New Scientist. . Retrieved 2008-06-21.
[6] Leonard David. "Virgin Galactic Spaceliner Steps Forward" (http://www.space.com/spacenews/businessmonday_070226.html). Space News. . Retrieved 2007-04-06.
[7] "Space tourism company to fly in 2008" (http://space.newscientist.com/article/dn7747). AirVenture Museum. . Retrieved 2007-04-06.
[8] "SpaceShipThree poised to follow if SS2 succeeds" (http://www.flightglobal.com/Articles/2005/08/23/Navigation/200/201097/SpaceShipThree+poised+to+follow+if+SS2+succeeds.html). Flight International. 23 August 2005. . Retrieved 2007-04-06.
[9] jnhtx (87543) (July 30 2006). "Notes on Rutan presentations at EAA Oshkosh" (http://science.slashdot.org/comments.pl?sid=157540&threshold=1&commentsort=5&tid=160&mode=nested&cid=13205378). Slashdot. . Retrieved 2007-04-06.
[10] Jefferson, Catherine (2004). "First Private Manned Space Flight" (http://www.devsite.org/lab/spaceshipone/data.shtml). .
[11] "SpaceShipOne" (http://www.astronautix.com/craft/spaipone.htm). *Encyclopedia Astronautica*. .
[12] http://scaled.com/projects/tierone/data_sheets/html/ox_tank.htm
[13] http://scaled.com/projects/tierone/index.htm
[14] http://scaled.com/projects/tierone/data_sheets/PDF/SS1_flight_profile.pdf
[15] http://scaled.com/projects/tierone/data_sheets/html/images/lithograph---SS1_flight_profile.jpg
[16] http://www.SpaceDev.com
[17] http://www.bbc.co.uk/newsa/n5ctrl/events04/sci/spaceship1/nb_60km.ram
[18] http://youtube.com/watch?v=_YWrjWWC7EI
[19] http://www.wired.com/news/space/0,2697,64123,00.html
[20] http://registry.faa.gov/aircraftinquiry/NNum_Results.aspx?NNumbertxt=N328KF
[21] http://www.DesertTurtle.com
[22] http://www.DesertTurtle.com/Song.html
[23] http://www.DesertTurtle.com/Oshkosh.html
[24] http://patft.uspto.gov/netacgi/nph-Parser?Sect1=PTO2&Sect2=HITOFF&p=1&u=%2Fnetahtml%2FPTO%2Fsearch-bool.html&r=1&f=G&l=50&co1=AND&d=PTXT&s1=spaceship&s2=(%22christie+parker%22.ATT.)&OS=spaceship+AND+LREP/IU.S.

Mike Melvill

Michael Winston "Mike" Melvill

Melvill receiving the first commercial astronaut wings

Commercial Astronaut

Born	November 11, 1940 Johannesburg, South Africa
Other occupation	Test Pilot
Time in space	~10 mins
Selection	SpaceShipOne 2003
Missions	SpaceShipOne flight 15P, SpaceShipOne flight 16P

Michael Winston "Mike" Melvill (born November 11, 1940) is one of the test pilots for SpaceShipOne, the experimental spaceplane developed by Scaled Composites. Melvill piloted SpaceShipOne on its first flight past the edge of space, flight 15P on June 21, 2004, thus becoming the first commercial astronaut and the 434th person to go into space. He was also the pilot on SpaceShipOne's flight 16P, the first competitive flight in the Ansari X Prize competition.

Early life / Family

Melvill was born in Johannesburg in South Africa before his family moved to Kloof near Durban. He attended Highbury Preparatory School and Hilton College. Later on Melvill moved to the United States from England in February 1967 and became a U.S. citizen in 1972. Melvill is married; he and his wife Sally have two sons and four grandchildren.

Test pilot

Melvill met Burt Rutan when he showed Rutan the Rutan VariViggen he had built at his home. Melvill was, having retired in October 2007, the vice president and general manager of Scaled Composites, LLC; he has 25 years of experience as a test pilot, and has logged nearly 7600 hours in over 130 types of aircraft. A longtime acquaintance of Scaled Composites founder Burt Rutan, Melvill has flown the first flights of ten of Rutan's aircraft. Melvill also helped develop the release mechanism for SpaceShipOne.

Awards & achievements

The first "Commercial Astronaut" Michael Winston "Mike" Melvill, reviews the first private space flight with Scott Crossfield

Melvill is a Fellow of the Society of Experimental Test Pilots; in 1999, he received their Iven C. Kincheloe award for his work in high-altitude development testing Scaled's Model 281 Proteus. Melvill also built his own Long EZ aircraft, which he flew around the world with Dick Rutan in 1997. He is also the only person other than Dick Rutan or Jeana Yeager to have piloted the Voyager aircraft. Melvill is the sole or joint holder of nine FAI aviation world records in various categories.

Melvill was the pilot for SpaceShipOne flight 15P, SpaceShipOne's first spaceflight and the first privately-funded human spaceflight mission to reach space, on the morning of June 21, 2004. In a ceremony two hours after landing, Melvill was awarded his astronaut wings, specifically the FAA Commercial Astronaut badge, the first wings awarded for a non-government space program and the first for a spaceplane flight since the X-15 flights of the 1960s. In 2004 Melville received his second Iven C. Kincheloe Award for his work on the SpaceShipOne program.

External links

- Scaled Composites biography [1]
- Sport Aviation article by Melvill describing his round-the-world trip [2]
- FAI aviation world records held by Michael W. Melvill [3]
- Spacefacts biography of Mike Melvill [4]

References

[1] http://www.scaled.com/projects/tierone/melvill.htm
[2] http://www.earthrounders.com/melvill.html
[3] http://records.fai.org/pilot.asp?from=ga&id=707
[4] http://www.spacefacts.de/bios/astronauts/english/melvill_michael.htm

Space Adventures

Type	Private
Industry	Commercial Spaceflight
Founded	1998
Headquarters	Vienna, Virginia, USA
Products	space tourism
Website	www.spaceadventures.com [1]

Space Adventures, Ltd. is a Virginia, USA based space tourism company founded in 1998 by Eric C. Anderson. As of 2010, they offer zero-gravity atmospheric flights and orbital spaceflights.[2] They have announced future plans for sub-orbital and lunar spaceflights.

Background

Eric C. Anderson is the president and CEO of Space Adventures. He co-founded the company in 1998 with several other entrepreneurs from the aerospace, adventure travel and entertainment industries. The company is headquartered in Vienna, Virginia with an office in Moscow.

Space Adventures offers a variety of programs such as Orbital spaceflight missions to the International Space Station, Circumlunar missions around the moon, zero gravity flights, cosmonaut training programs, spaceflight qualification programs, and reservations on future suborbital spacecrafts.

Since 2001, Space Adventures has launched seven clients on eight successful missions to the International Space Station (ISS). In April 2001, the company sent American businessman Dennis Tito for a reported $20 million payment, making him the first space tourist. South African businessman Mark Shuttleworth did the same in April 2002, becoming the first African in space. Gregory Olsen became the third private citizen to travel to the ISS in October 2005, followed by the first female space tourist, Anousheh Ansari, who completed her 10-day orbital mission in September 2006. Charles Simonyi, an ex-executive at Microsoft, became the fifth space tourist who visited the ISS in April 2007, then again in March 2009. He is the world's first private space explorer who launched to space twice. In 2008, game developer Richard Garriott, the first second-generation U.S. astronaut, became the sixth client to travel to the ISS. In October 2009, Cirque du Soleil founder, Guy Laliberté, became the first Canadian space tourist to have launched to space.

In January 2008, Space Adventures acquired Zero Gravity Corporation which is the first and only FAA-approved provider of weightless flights to the general public.

The company's advisory board includes Apollo 11 moonwalker Buzz Aldrin, Space Shuttle astronauts Sam Durrance, Tom Jones, Byron Lichtenberg, Norm Thagard, Kathy Thornton, Pierre Thuot, and Charles Walker, Skylab/Shuttle astronaut Owen Garriott, and Russian cosmonaut Yuri Usachev.[3]

Clients

Mark Shuttleworth in space

Dennis Tito

Space Adventures' first orbital spaceflight client and the world's first private space explorer launched to the ISS in April 2001 on Soyuz TM-32. American businessman Dennis Tito received training at the Yuri Gagarin Cosmonaut Training Center in Star City in Russia. Tito participated in Space Adventures' other programs, including a Zero-Gravity flight, centrifuge training and a supersonic jet flight to the Edge of Space prior to his orbital flight.

Mark Shuttleworth

Mark Shuttleworth spent 10 days in space. He launched with two crewmates, Russian commander Yuri Gidzenko and Italian astronaut Roberto Vittori. They launched on Soyuz rocket TM-34 from the Baikonur Cosmodrome, Kazakhstan on April 25, 2002. Prior to his flight, Mr. Shuttleworth completed Space Adventures' Orbital Pre-Qualification Program and underwent almost eight months of training and medical exams, including a one-week orientation program at NASA's Johnson Space Center in Houston. A zero-gravity flight, centrifuge training, and spacecraft communication, guidance and control system lessons of the Russian Soyuz spacecraft and ISS were also part of his training. Shuttleworth dedicated his flight to educating South African youth and conducting scientific research.

Gregory Olsen

Dr. Olsen completed over 900 hours of training in Star City, Russia in preparation for his mission. He and his crew mates launched from the Baikonur Cosmodrome on October 1, 2005 aboard Soyuz TMA-7. While aboard the ISS, he participated in a research program prepared by the European Space Agency that studied the human body's response to the microgravity environment. Through Amateur Radio on the ISS, Dr. Olsen contacted high school students in New Jersey and New York.[4]

Anousheh Ansari

Anousheh Ansari lifted off on Soyuz TMA-9 on September 18, 2006 from Baikonur, Kazakhstan. Ansari became the fourth (and first female) space tourist. During her eight day stay on board the International Space Station, Ansari conducted four experiments for the European Space Agency, including researching the mechanisms behind anemia, how changes in muscles influence lower back pain, consequences of space radiation on ISS crew members and different species of microbes that have made a home for themselves on the space station.[7]

Charles Simonyi

Dr. Simonyi is the first repeat orbital spaceflight client of Space Adventures'. His first spaceflight mission was in 2007 aboard Soyuz TMA-10 and his second was in 2009 aboard Soyuz TMA-14. Simonyi's goals for both of his missions were to advance civilian spaceflight, assist space station research and involve the world's youth in the science of space travel. His website, www.charlesinspace.com, had 33 million viewers.[5]

Richard Garriott

Garriott became the first second-generation space traveler following his astronaut father Owen Garriott into space. He is also the second person to wear the British flag in space. He launched for the International Space Station on October 12, 2008 aboard Soyuz TMA-13. Richard's main objective for his mission was to encourage commercial participation. By fostering the involvement of individuals, companies and organizations in his spaceflight, Richard hoped to demonstrate that there is commercial potential in private space exploration, while furthering the understanding of space.

Guy Laliberté

Laliberté is the first Canadian space tourist who reached orbit on September 30, 2009 aboard Soyuz TMA-16. While in orbit, Laliberté promoted the One Drop Foundation and proclaimed his mission as a "Poetic Social Mission". He also conducted the first ever artistic and social event, "Moving Stars and Earth for Water," to originate from space that took place on October 9, 2009.[9] It was a two hour event that was hosted by Laliberté and many celebrities such as Salma Hayak, Shakira and Bono participated from Earth.[6] He returned to Earth on board Soyuz TMA-14.

Orbital Mission Explorers Circle

This program allows individuals to reserve seats on future orbital spaceflights. They have the option to fly to orbit as their schedule allows with preferential access to mission seats or they can opt to sell their seat to another private astronaut. Sergey Brin, co-founder and president of technology for Google Inc., has become the founding member of the "Founding Explorer" group by placing a $5 million deposit on a future orbital spaceflight.

Space Adventures has established the Orbital Mission Explorers Circle to build a definitive consortium of future private space explorers who share a lifetime goal of orbital spaceflight or the investment therein. Space Adventures has initially created six "Founding Explorer" positions in the Orbital Mission Explorers Circle, each of whom will have priority access to participate in future orbital space missions.

Spacewalk

In 2006, the company announced that it would begin offering a spacewalk option to its clients traveling to the ISS. The spacewalk would allow participants to spend up to 1.5 hours outside of the space station and costs about $15 million. It would lengthen the orbital mission approximately six to eight days. The spacewalk would be completed in the Russian designed Orlan space suit. The training for the spacewalk would require an extra month of training on top of the six months already required.[7]

Lunar Mission

Space Adventures is offering advance booking for a future lunar mission involving travel to circumnavigate the moon, on a circumlunar trajectory. Pricing has been announced at US$100 million per seat. This mission will utilize two Russian launch vehicles. One Soyuz capsule will be launched into low-earth orbit by a Soyuz rocket. Once in orbit, the manned capsule will dock with a second, unmanned, lunar-propulsion module which will then power the circumlunar portion of the trip.[8] The mission will last 8–9 days. It will take 2–3 days to approach the Upper Stage (Block DM), 2–3 days to approach the moon, there's approximately a 45 minute observation of the moon from

100 km-1000 km above the moon's surface, and will return to Earth in 2–3 days. No passengers have yet been carried.

Suborbital initiatives

Explorer suborbital vehicle

The company, along with Prodea and FSA, began to develop a suborbital space transportation system, called Explorer. The vehicle has been designed by Myasishchev Design Bureau, a Russian aerospace organization which has developed a other aircraft and space systems.

The Explorer aerospace system will consist of a flight-operational carrier aircraft, the M-55X, and a rocket spacecraft. It will have the capacity to transport up to five people to space.[9]

Space Adventures abandoned the Explorer project because "it got too expensive."[10]

Armadillo suborbital rocket

Armadillo Aerospace is currently working on a sub-orbital commercial rocket that will be marketed through Space Adventures. The per-passenger price point was announced in April 2010 to be $102,000 for a flight to 62 miles altitude, above the Karman line.[11]

The Armadillo-developed technology will be a vertical takeoff, vertical landing (VTVL) suborbital vehicle carrying space tourists to at least 100 kilometers altitude, with Space Adventures selling the seats. The spacecraft development effort is jointly funded by both Armadillo and Space Adventures.[10]

Lawsuit

Japanese businessman Daisuke Enomoto sued Space Adventures as his trip was canceled by them for medical reasons after he paid $21,000,000 and no refund was given.[12] In June 2009, the lawsuit was dismissed.

See also

- Space tourism
- Space colonization
- Private spaceflight
- Commercial Astronaut
- Space Tourism Society
- List of private spaceflight companies
- Quasi Universal Intergalactic Denomination

External links

- Space Adventures [1]
- Space Daily article on 2004/2005 space tourist contract [13]
- Spaceport Singapore [14]
- Space tourism: ready for the masses? [15] 20 June 2006

References

[1] http://www.spaceadventures.com/

[2] SpaceAdventures.com (http://www.spaceadventures.com/)

[3] Kim Hart (2007-04-21). "Travel agency launches tourists on out-of-this-world adventures" (http://seattletimes.nwsource.com/html/businesstechnology/2003675354_spacetours21.html?syndication=rss). Seattle Times. . Retrieved 2008-05-26.

[4] http://www.comspacewatch.com/news/viewpr.html?pid=18011

[5] http://www.redorbit.com/news/space/910069/space_adventures_client_charles_simonyi_returned_to_earth_after_completing/index.html

[6] http://www.parabolicarc.com/2009/09/07/full-details-guy-lalibertes-space-tourism-flight/#more-8665

[7] Tariq Malik (2006-06-21). "Space Adventures Offers $15 Million Spacewalks for ISS Visitors" (http://www.space.com/news/060721_spacetourist_eva.html). Space.com. . Retrieved 2008-05-26.

[8] Lunar Mission (http://spaceadventures.com/index.cfm?fuseaction=Lunar.Details), *SpaceAdventures website, undated, accessed 2010-05-24.*

[9] Alan Boyle (2006-02-17). "New group to develop passenger spaceship" (http://www.msnbc.msn.com/id/11393569/). MSNBC. . Retrieved 2008-05-26.

[10] Space Adventures returns to suborbital spaceflight (http://www.newspacejournal.com/2010/05/28/space-adventures-returns-to-suborbital-spaceflight/), *NewSpace Journal*, 2010-05-28, accessed 2010-05-28.

[11] Perhaps You'll Visit Space In Your Lifetime, After All (http://gizmodo.com/5537278/), Gizmodo, 2010-05-12, accessed 2010-05-15.

[12] "'Japanese businessman defrauded of $21 million by the U.S. space tourism company'" (http://www.japanprobe.com/?p=6351). Japan Probe. 2008-09-26. . Retrieved 2008-09-28.

[13] http://www.spacedaily.com/news/tourism-03d.html

[14] http://www.spaceportsingapore.com/

[15] http://www.janes.com/aerospace/civil/news/misc/janes060620_1_n.shtml

Virgin Galactic

Type	Private
Industry	Space tourism
Founded	2004
Headquarters	New Mexico, USA
Key people	Richard Branson (UK) Will Whitehorn George Whitesides
Owner(s)	Virgin Group
Website	www.virgingalactic.com [1]

Virgin Galactic is a company within Sir Richard Branson's Virgin Group which plans to provide sub-orbital spaceflights to the paying public, along with suborbital space science missions and orbital launches of small satellites. Further in the future Virgin Galactic hopes to offer orbital human spaceflights as well.

Spacecraft

Overview of the spacecraft flights

SpaceShipOne, the first private manned spacecraft to fly above the 100-km Kármán line

It is planned that the spacecraft are to be robust and affordable enough to take paying passengers ($200,000 per person with a $20,000 deposit). The craft is projected to be a six passenger, two pilot craft.[2] Its planned trajectory will overlap the Earth's atmosphere at 70,000 feet (21,000 m), which will make it a sub-orbital journey with a short period of weightlessness. At around 16 kilometers or 52000 ft (16000 m), the rocket will separate when the plane reaches its maximum height. The time from liftoff of the White Knight II booster carrying SpaceShipTwo until the touchdown of SpaceShipTwo after the sub-orbital flight will be about 2.5 hours. The sub-orbital flight itself will only be a small fraction of that time. The weightlessness will last approximately 6 minutes.[3] Passengers will be able to release themselves from their seats during these 6 minutes and float around the cabin.[2]

In addition to the sub-orbital passenger business, Virgin Galactic will market SpaceShipTwo for sub-orbital space science missions and market WhiteKnightTwo for "small satellite" launch services. They plan to initiate RFPs for the satellite business in early 2010.[4]

SpaceShipTwo's performance

WhiteKnight2

SpaceShipTwo will fly higher than SpaceShipOne, to a height of 110 km in order to go beyond the defined boundary of space (100 km) and lengthen the experience of weightlessness.[2] The spacecraft will reach around Mach 3 (1000 m/s), which is faster than current fighter jets are capable of attaining; however, the spacecraft will not be able to sustain that speed for long periods of time. It has double the crew (2) and can carry triple the passengers (6) of its predecessor. In honour of the science fiction series "Star Trek", the first two ships are named in honour of the fictional starships "Enterprise" and "Voyager". To re-enter the atmosphere SpaceShipTwo folds its wings up, and then returns them to their original position for a smooth and gentle glide back onto the runway. The craft has a very limited cross-range capability and until other planned spaceports are built worldwide, it has to land in the area where it started. Further "Spaceports" are planned in Dubai and elsewhere, with the intention that the spaceline will have a world-wide availability and commodity in the future.

SpaceShipTwo

Sir Richard Branson unveiled the rocket plane on Monday 7th December 2009. SpaceShipTwo was presented to the world in the Mojave desert, in California. The vehicle will undergo testing over the next 18 months before being allowed to take ticketed individuals on short-hop trips just above the atmosphere. Sir Richard, who heads the Virgin Group, intends to run the first flights out of New Mexico before extending operations around the globe. Built from lightweight carbon composite materials and powered by a hybrid rocket motor, SS2 is based on the X-Prize-winning SpaceShipOne concept - a rocket plane that is lifted initially by a carrier vehicle before blasting skywards. SS1 became the world's first private spaceship with a series of high-altitude flights in 2004.

Its successor, however, is twice as large, measuring 18m (60ft) in length. And whereas SpaceShipOne only had a single pilot (and the ballast equivalent of two passengers), SS2 will have a crew of two and room for six passengers. About 300 individuals are reported to have signed up for a flight. They are all willing to pay about $200,000 (£121,000) for the privilege of experiencing six minutes of weightlessness during what will be a two-hour end-to-end flight. Many of those future "astronauts" have attended the VIP unveiling at the Mojave Air and Spaceport. California Governor Arnold Schwarzenegger and New Mexico Governor Bill Richardson christened the spaceliner Virgin SpaceShip (VSS) Enterprise. The New Mexico authorities are investing almost $200m (£121m) in a purpose-built facility in Upham. It will have a 3,000m (10,000ft) runway and a suitably space-age terminal and hangar building designed by Foster and Partners. Sir Richard's Virgin Galactic enterprise will have competitors but he is almost certain to be the first to market, barring any problems arising in the test campaign. SpaceShipTwo's carrier plane is called WhiteKnightTwo. It was finished last year and has already begun its own trials.

The Spaceship Company

The Spaceship Company (TSC) is a new aerospace production company, founded by Sir Richard Branson's Virgin Group and Scaled Composites, which is building a fleet of commercial spaceships and launch aircraft with the intention of making widespread space travel a reality. TSC's initial launch customer is Virgin Galactic, which has contracted to purchase five SpaceShipTwos and two WhiteKnightTwos.[5] To meet Virgin Galactic's requirements, TSC has contracted Scaled Composites to develop and build prototypes of WK2 and SS2, which TSC started full-scale production of in 2008.[6] [7]

Competition

There are numerous other companies actively working on commercial passenger suborbital spaceflight. Additionally, there are several others developing commercial manned orbital spaceflight capability (including some which are initially designed for, or may eventually be used for, commercial passenger spaceflight), which is a significantly more difficult problem than suborbital spaceflight.

Fleet

As of 2008, Virgin Galactic is planning to have a fleet of two White Knight Two motherships and five or more SpaceShipTwo tourist suborbital spacecraft.[8]

WhiteKnightTwo carrier aircraft

Virgin MotherShips

1. *VMS Eve* [9]
2. *VMS Spirit of Steve Fossett* [10] [11] [12]

SpaceShipTwo suborbital spacecraft

Virgin SpaceShips

1. VSS *Enterprise* [13]
2. VSS *Voyager* [14] [15]
3. Not Yet Named
4. Not Yet Named
5. Not Yet Named

Base

Test launches are planned to take place from the Mojave Spaceport, where Scaled Composites is constructing the spacecraft. Virgin Galactic expects that initial passenger flights will take place there, as well.[16]

An international architectural competition was held for the design of Virgin Galactic's operating base, Spaceport America in New Mexico. The contract was awarded to URS and Foster + Partners architects.[17] Upon the completion of the spaceport, flight operations will transfer from the California desert to the new bases.[16]

Virgin Galactic also announced on April 3, 2008 that in the future they will operate in Europe out of Spaceport Sweden.[18]

Will Whitehorn stated in an interview with Space.co.uk on 29 April 2008 that the company was considering flying from a UK base: RAF Lossiemouth in Scotland, during the summer months only.[19] There are also rumours that the base could be located in RAF Machrihanish (Campbeltown, Scotland) or an unnamed location in south-west England.[20]

Commencement of space flights

Will Whitehorn, president of Virgin Galactic, has stated that the company would "not put a definite timeline on when the commercial flights would begin" but that "all was on track with its development plans" and that "If all goes to plan", the inaugural sub-orbital flight should happen "within two years [of June 2009]".[21]

Refining the projected schedule in October 2009, Virgin Galactic has continued to decline to announce a firm schedule for commercial flights, but did reiterate that initial flights would take place from Spaceport America. Commercial availability will be based on a "safety-driven schedule," which they hope to achieve "within two years."[4]

Scaled Composite's president Doug Shane has said WhiteKnight Two's first SpaceShipTwo captive flights will be in early 2010.[22] Both aircraft flew together in March 2010 [23] . On December 7, 2009, SpaceShipTwo was unveiled at the Mojave Spaceport.[24]

Collaboration with NASA

In February 2007, officials from Virgin and NASA signed a memorandum of understanding to explore the potential for collaboration. [25] [26]

See also

- Commercial Astronaut
- List of private spaceflight companies
- Space Adventures
- Space colonization
- Space tourism
- Space Tourism Society
- X Prize Foundation

External links

- The Spaceship Company [27]
- Virgin Galactic's SpaceShipTwo Mothership Makes Maiden Flight [28]
- Virgin Galactic [1]
- Virgin Galactic:Let the Journey Begin (Video) [29]
- Branson And Rutan Launch New Spaceship Manufacturing Company [30]
- U.S. Okays Virgin Galactic Spaceship Plans [31]
- New Mexico Spaceport Bills Signed [32]
- Lloyds Eyes Covering Virgin Spaceflights [33]
- The Future of Travel [34]
- Virgin Galactic Rolls Out Mothership "Eve“ [35]

References

[1] http://www.virgingalactic.com

[2] "Virgin Galactic" (http://www.virgingalactic.com/htmlsite/faq.php). virgingalactic.com. . Retrieved 2008-02-10.

[3] "Captain Kirk signs on for Virgin Galactic Space Ride" (http://www.soultek.com/space_privatization/Virgin_Galactic_Captain_Kirk_and_spaceshipone.htm). soultek.com. . Retrieved 2008-02-14.

[4] Will Whitehorn. (2009-10-27). *Internatonal Astronautical Congress 2009: Civilian Access to Space* (http://www.flightglobal.com/blogs/hyperbola/2009/10/video-iac2009-civilian-access.html). [video, comments at c. 22:00]. Daejeon, Korea: Flightglobal Hyperbola, Rob Coppinger. .

[5] "Richard Branson and Burt Rutan Form Spacecraft Building Company" (http://www.space.com/news/050727_branson_rutan.html). SPACE.com. 2005-07-27. . Retrieved 2009-10-17.

[6] http://www.virgingalactic.com/jobs/Vehicle%20Production%20Planner.pdf

[7] http://www.scaled.com/projects/ttop/press_release.pdf

[8] Malik, Tariq (2008-01-23). "Virgin Galactic Unveils Suborbital Spaceliner Design" (http://www.space.com/news/080123-virgingalactic-ss2-design.html). *SPACE.com*. . Retrieved 2008-01-25.

[9] "Spaceship Company unveils design of SpaceShipTwo" (http://newsfromrussia.com/news/science/23-01-2008/103605-spacecraft-0). *Pravda Online*. 2008-01-23. . Retrieved 2008-01-25.

[10] Branson, Richard (10 October 2007). "My Friend, Steve Fossett" (http://www.time.com/time/nation/article/0,8599,1670216,00.html). *Time*. . Retrieved 2007-10-31.

[11] Burack, Ari (10 October 2007). "Sir Richard Branson, black robed as Father Richard for zany party inaugurating Virgin American flight from San Francisco to Las Vegas" (http://www.sanfranciscosentinel.com/?p=5903). *San Francisco Sentinel*. . Retrieved 2008-02-28.

[12] Nizza, Mike (2007-10-11). "The Legend of Steve Fossett Takes Root" (http://thelede.blogs.nytimes.com/2007/10/11/the-legend-of-steve-fossett-takes-root/index.html?ex=1349841600&en=45eab0da4e07992c&ei=5088&partner=rssnyt&emc=rss). *New York Times*. . Retrieved 2007-12-20.

[13] "Virgin boss in space tourism bid" (http://news.bbc.co.uk/2/hi/science/nature/3693020.stm). BBC News. 27 September 2004. . Retrieved 2007-10-31.

[14] Wilson, Simon (2005-01-21). "Will space tourism ever take off?" (http://www.moneyweek.com/file/2333/space-tourism.html). *MoneyWeek*. . Retrieved 2007-12-20.

[15] "One small step for space tourism... Private spaceflight" (http://economist.com/science/displaystory.cfm?story_id=E1_PVDDNPS&CFID=652299&CFTOKEN=40feddda82a83b6-F677FAB6-B27C-BB00-0143FB2BB9733AFB). *The Economist*. 16 December 2004. . Retrieved 2007-12-20.

[16] "Virgin Galactic FAQ: Where Will I Fly From?" (http://www.virgingalactic.com/htmlsite/overview.php). . Retrieved 2008-06-26.

[17] "Foster + Partners" (http://www.fosterandpartners.com/Projects/1613/Default.aspx). fosterandpartners.com. . Retrieved 2008-02-17.

[18] "News Release 03.04.2008 / Spaceport Sweden and Virgin Galactic" (http://www.virgingalactic.com/htmlsite/news.php). . Retrieved 2008-06-26.

[19] "Will Whitehorn (Virgin Galactic) and Heather MacRae (Venture Thinking) at the RAeS" (http://www.space.co.uk/DataBank/VideoGallery/VideoPlayer/tabid/384/VideoId/37/Will-Whitehorn-Virgin-Galactic-And-Heather-MacRae-Venture-Thinking-At-The-RAeS.aspx). space.co.uk. . Retrieved 2008-07-26.

[20] "SNP backs Lossiemouth as first UK space port" (http://www.techradar.com/news/world-of-tech/snp-backs-lossiemouth-as-first-uk-space-port-497157). . Retrieved 2008-12-26.

[21] "Work starts on New Mexico spaceport" (http://news.bbc.co.uk/1/hi/sci/tech/8111243.stm). *BBC News*. 2009-06-20. . Retrieved 2010-05-27.

[22] http://www.flightglobal.com/articles/2009/10/22/333823/pictures-whiteknight-twos-spoilers-get-holes.html

[23] http://www.virgingalactic.com/news/item/vss-enterprises-first-flight/

[24] Allen, Nick (2009-12-08). "Richard Branson unveils Virgin Galactic's spaceship Enterprise" (http://www.telegraph.co.uk/travel/travelnews/6761213/Richard-Branson-unveils-Virgin-Galactics-spaceship-Enterprise.html). *The Daily Telegraph* (London). . Retrieved 2010-05-27.

[25] http://www.nasa.gov/pdf/169943main_NASA_Virgin_MOU.pdf

[26] http://www.spaceref.com/news/viewpr.html?pid=21973

[27] http://www.thespaceshipcompany.com

[28] http://spacefellowship.com/News/?p=7772

[29] http://www.space.com/php/video/player.php?video_id=Virgin_galactic01

[30] http://www.spacedaily.com/news/spacetravel-05zzzg.html

[31] http://www.space.com/news/050815_virgingalactic_itar.html

[32] http://www.msnbc.msn.com/id/11630251

[33] http://www.space-travel.com/reports/Lloyds_Eyes_Covering_Virgin_Spaceflights.html

[34] http://www.shavemagazine.com/lifestyle/090302/3

[35] http://spacefellowship.com/News/?p=6155

Bigelow Aerospace

Type	Private
Industry	Aerospace
Founded	1999[1]
Founder(s)	Robert Bigelow (Founder and President)
Headquarters	North Las Vegas, Nevada, United States
Products	Orbital facilities, commercial space stations
Employees	120 (2006)
Website	BigelowAerospace.com [2]

Bigelow Aerospace is a North Las Vegas, Nevada space technology startup company that is pioneering work on expandable space station modules. Bigelow Aerospace was founded by Robert Bigelow in 1998[3] and is funded in large part by the fortune Bigelow gained through his ownership of the hotel chain Budget Suites of America. By 2010, Bigelow had invested US$180 million in the company.[4] Bigelow has stated on multiple occasions that he is prepared to fund Bigelow Aerospace with about US$500 million through 2015 in order to achieve launch of full-scale hardware.[3] [5]

History

Bigelow originally licensed the multi-layer, expandable space module technology from NASA after NASA cancelled the ISS TransHab project due to budget constraints in the late 1990s.[6] Bigelow continued to develop the technology for a decade, redesigning the module fabric layers -- including adding proprietary extensions of Vectran shield fabric, "a double-strength variant of Kevlar" -- and developing a family of uncrewed and crewed expandable spacecraft in a variety of sizes.[6] . Bigelow invested US$75 million in proprietary extensions to the NASA technology by mid-2006 ,[1] and $180 million into the technology by 2010.[4] In early 2010, NASA came full circle to once again investigate "making inflatable space-station modules to make roomier, lighter, cheaper-to-launch spacecraft" by announcing plans in its budget proposal released 22 February 2010. NASA is "considering connecting a Bigelow expandable craft to the ISS to verify their safety by testing life support, radiation shielding, thermal control and communications capabilities."[6]

Since early on, Bigelow has been intent on "pursuing markets for a variety of users including biotech and pharmaceutical company and university research, entertainment applications and government military and civil users." The business model includes "'leasing out' small space stations or habitats made of one or more Nautilus inflatable modules to different research communities or corporations."[3] . Despite these broad plans for space commercialization, the *space tourism destination* and *"space hotel"* monikers were frequently used by many media outlets following the 2006/2007 launches of Genesis I and Genesis II. Robert Bigelow has been explicit that he is aiming to do business in space in a new way, with "low cost and rapid turnaround, contrary to traditional NASA ISS and space shuttle operations and bureaucracy."[3]

Module design and business plans

An expandable module is a space structure that has a flexible outer shell, allowing conservation of diameter for launch and weight overall. Once in orbit, the module is inflated to allow for greater work, play and living area for astronauts. Expandable modules were initially proposed and designed by the United States National Aeronautics and Space Administration (NASA) under the TransHab program. After cancellation of the Transhab program, Bigelow Aerospace entered into three Space Act agreements whereby Bigelow Aerospace is the sole commercializer of several of NASA's key expandable module technologies.

The company is developing a family of prototype and production space station modules, including:

Model of the proposed BA 330 module.

- Genesis I, a one-third scale prototype module launched on July 12, 2006 weighing approximately 3,000 pounds (1,360 kg), with approximate dimensions of 14.5 feet (4.4 meters) in length and 5.25 feet (1.6 meters) in diameter, expanding to nearly twice the diameter once in orbit;[7]
- Genesis II, with the same dimensions as Genesis I but with additional systems to be tested and items being sent up by paying members of the public. It was launched on June 28, 2007 .[8]
- Sundancer, the first human inhabited test module
- BA 330, a full-scale production module weighing approximately 50,000 pounds (23,000 kg) when fully loaded, with dimensions of approximately 45 feet (13.7 meters) in length and 22 feet (6.7 meters) in diameter when expanded.[9] The BA 330 was formerly referred to as the Nautilus.

Bigelow Aerospace anticipates that its inflatable modules will be more durable than rigid modules.[10] This is partially due to the company's use of several layers of vectran, a material twice as strong as kevlar, and also because, in theory, flexible walls should be able to sustain micrometeorite impacts better than rigid walls. In ground-based testing, micrometeoroids that were capable of puncturing standard ISS module materials only penetrated about half-way through the Bigelow skin. Operations director Mike Gold commented that Bigelow modules also wouldn't suffer from the same local shattering problems likely with metallic modules. This could result in as much as 24 hours to remedy punctures in comparison to the more serious results of standard ISS skin micrometeoroid damage. [6]

Expected uses for Bigelow Aerospace's expandable modules include microgravity research and development and space manufacturing. Other potential uses include space tourism, such as modules for orbital hotels, and space transportation, such as components in spaceships for Moon or Mars manned missions. The company plans to sell BA 330 modules for US$100 million apiece. Bigelow also plans to launch by 2010 an orbital resort, tentatively called the CSS Skywalker.

On April 10, 2007, Bigelow Aerospace announced business plans to offer (by 2012) a four-week orbital stay for US$15 million, with another four weeks for an additional US$3 million. An entire orbital facility could also be leased for US$88 million a year, or half a facility for US$54 million a year.[11] Speaking during a teleconference organised by the Commercial Spaceflight Federation in February 2010 Bigelow announced a new price of $23 million for a 30 day stay on a Bigelow habitat. That price covers everything including transport, training, and consumables.[12]

In 2010, Bigelow proposed conceptual designs for expandable habitats that would be substantially larger than the BA-330, previously its largest at 330 m^3 habitat volume. Contingent on NASA going forward with a super heavy lifter, the proposed concept would include "expandable habitats offering 2,100 cubic meters of volume — nearly twice the capacity available on the International Space Station" and another providing 3,240 cubic meters.[13]

CSS Skywalker

CSS Skywalker		
Description		
Role:	Space Hotel	
Crew:	5 - 7	
Dimensions		
Height:	98.0 ft	30.0 m
Diameter:	21.9 ft	6.7 m
Habitable Volume:	52,972 ft^3	1,500 m^3
Masses		
Total:	220,000 lb	100,000 kg
Electrical System		
Solar Cells		
CSS Skywalker		

The **CSS Skywalker** (**Commercial Space Station Skywalker**) is a 2005 concept for the first space hotel by Bigelow Aerospace.[14] The Skywalker was designed to be composed of multiple Nautilus habitat modules, which would be inflated and connected upon reaching orbit. An MDPM (Multi-Directional Propulsion Module) would allow the Skywalker to be moved into interplanetary or lunar trajectories.[15] As of 2009, Bigelow is continuing to develop a variety of space hotel architectures.[16]

Module construction and deployment timeline

On July 12, 2006 and June 28, 2007, Bigelow launched the Genesis I and II modules, respectively. In mid-2008, Bigelow Aerospace completed the Galaxy module but did not launch it due to rising launch costs[17] and, the ability to substantially validate the new Galaxy technologies terrestrially, particularly after the successful two Genesis launches in 2006 and 2007.[18] It was tested on the ground at its North Las Vegas facility instead.

Bigelow has reserved a 2014 launch on the SpaceX Falcon 9 rocket[19] , but has not yet announced the payload. The Falcon 9 is capable of launching either a Sundancer or a BA 330 module. Bigelow is also in talks with Lockheed Martin to potentially contract launch services on its Atlas V - 401 vehicle.[20] [21]

Note: Dates of upcoming launches are proposed and are subject to change.

Module Type	Module Name	Flight Date	Launch Vehicle	Status
Genesis Pathfinder	Genesis I	July 12, 2006 14:53 UTC	Dnepr	Launch successful, on orbit[22]
Genesis Pathfinder	Genesis II	June 28, 2007 15:02 UTC	Dnepr	Launch successful, on orbit[23]
Galaxy	Galaxy	Canceled	none	Launch canceled, tests on ground [18]
Sundancer	unknown	unknown	unknown	Under construction[6]
BA 330	unknown	unknown	unknown	In design; mockup built

Genesis I

Genesis I, the first Bigelow Aerospace module to be placed into orbit

On July 12, 2006, Genesis I launched on a Dnepr booster from Yasny Launch Base in Siberia. The launch was conducted by Bigelow and ISC Kosmotras. Despite ground-side difficulties during launch, the spacecraft performed as expected upon reaching orbit, inflating, deploying solar arrays and starting internal systems.[24] The mission is planned to last for five years and include extensive observation of the craft's performance including testing packing/deployment procedures and resistance to radiation and space debris, among other space hazards and conditions. Mike Gold, corporate counsel for Bigelow Aerospace, stated in relation to this mission and the next, "Our motto at Bigelow Aerospace is 'fly early and often'. Regardless of the results of Genesis 1, we will launch a follow-up mission rapidly."[22]

Genesis II

On June 28, 2007, Genesis II launched on another Dnepr (a converted SS-18 ICBM) from Yasny Launch Base in Siberia. Launched at 8:02 a.m. PDT Genesis II was inserted into orbit at 8:16 a.m. PDT at an inclination of 64 degrees.

Although Genesis I and Genesis II are identical in size and similar in appearance there are several notable differences. Firstly, Genesis I contains 13 video cameras whereas Genesis II contains 22. Secondly, Genesis II includes a suite of additional sensors and avionics that are not present in Genesis I.

Fly your stuff program

Bigelow Aerospace ran a Fly Your Stuff [25] program for the Genesis II launch. The cost to launch pictures or small items was around US$300. Bigelow now photographs each item with internal cameras as the items float inside the craft, displaying them on the company website.

The first image of the interior of Genesis II appeared on the company's website on June 29, 2007. Some of the pictures and other items placed aboard Genesis II as part of the Fly Your Stuff program are clearly visible. Another interior image, apparently taken with more of the spacecraft's internal lights activated, was posted on July 2, 2007. Articles from the Fly Your Stuff program are also visible in this image.

Test items, supplied by Bigelow Aerospace employees, were sent into orbit on Genesis I. No new images of items floating inside Genesis I have been released since shortly after the launch and initial activation of the spacecraft due to problems with a computer which controls several of the internal cameras.[26]

Sundancer

The currently third planned Bigelow launch, *Sundancer*, will be equipped with full life support systems, attitude control, orbital maneuvering systems, and will be capable of reboost and deorbit burns.[27] Like the *Genesis* pathfinders, *Sundancer* will launch with its outer surface compacted around its central core, with air expanding it to its full size after entering orbit. After expansion, the module will measure 8.7 m (28.5 ft) in length and 6.3 m (20.6 ft) in diameter, with 180 cubic meters (greater than 6,000 cubic feet) of interior volume.[28] Unlike previous Bigelow craft, it will feature three observation windows.[27] As of September 2009, SpaceX had been contracted to provide a Falcon 9 vehicle for launch of a Bigelow payload in 2011.[29] However, as of February 2010 the only slot reserved for Bigelow on the SpaceX launch manifest is in 2014, and that manifest makes no claim about the nature of the Bigelow payload (Sundancer, or BA330, or otherwise).[19]

Crew and passenger transport

Bigelow's business model requires a means of transporting humans to and from low earth orbit. In 2004 Bigelow established and funded a $50 million prize, America's Space Prize, to stimulate development of manned vehicles. The prize expired without a winner in early 2010.

In August, 2009, Bigelow Aerospace announced the development of the Orion Lite spacecraft, intended to be a lower cost, and less capable version of the Orion spacecraft under development by NASA. The intention would be for Orion Lite to provide access to low earth orbit using either the Atlas 5 or Falcon 9 launch systems, and carrying a crew of up to 7.[30]

At the time Bigelow Aerospace's corporate counsel Mike Gold said: "...we would be foolish to depend completely on one capsule provider or any single launch system," ..."Therefore, it is vital from both a practical and business perspective to ensure that SpaceX and Dragon aren't the only options available to us, hence the need for another capsule."[31]

Bigelow entered NASA's Commercial Crew Development (CCDev) program with one proposal of its own and another one in collaboration with Boeing. Since then NASA has awarded Boeing $18 million for initial development of a crew capsule as part of CCDev. Bigelow is working with Boeing to refine requirements.

Aspirations beyond Earth-orbit

In February, 2010, following the announcement of NASA's post-Augustine Commission plans to reorient human-to-orbit plans more in the direction of commercial launch providers, Robert Bigelow said "We as a company have lunar ambitions. ... and we also have Mars ambitions as well."[32]

In April 2010, Mr Bigelow talked about having a space station at Lagrange point L1 and said his proposed private Moon Base would consist of three BA 330s.[13]

Honors

Bigelow Aerospace has received several honors for its spaceflight efforts. On October 3, 2006, Bigelow Aerospace received the Innovator Award from the Arthur C. Clarke Foundation.[33] The award recognizes "initiatives or new inventions that have had recent impact on or hold particular promise for satellite communications and society, and stand as distinguished examples of innovative thinking." Robert Bigelow was presented the award at the Arthur C. Clarke Awards in Washington D.C. alongside Walter Cronkite, who was honored on the same night with the Arthur C. Clarke Lifetime Achievement Award.

On January 26, 2007, the Space Foundation announced that Bigelow Aerospace would be the recipient of its 2007 Space Achievement Award.[34] Bigelow Aerospace joins a list of previous winners that include the Titan Launch Vehicle team; The Inertial Upper Stage team, the SpaceShipOne team; the Arianespace-CNES Ariane 4 launch team; the Evolved Expendable Launch Vehicle (EELV) teams; the NASA/Industry Galileo space probe team; the Hubble Space Telescope team; Sea Launch; and the NASA/Boeing International Space Station team. The award was presented to Robert Bigelow on April 9, 2007 at the 23rd National Space Symposium in Colorado Springs, Colorado.

See also

- Commercial Spaceflight Federation
- Blue Origin
- List of private spaceflight companies
- Space Adventures
- Space tourism
- Space architecture
- Virgin Galactic
- Transhab

External links

- Inflatable POOFs (privately owned orbital facility) [35], Space Review article (July 19, 2004)
- Bigelow Aerospace SpaceFellowship Archive [36], Space Fellowship Bigelow Aerospace News archive
- U.S. Hotel Tycoon Reaches for the Stars [37], Reuters article (August 8, 2004)
- Holidays in space are on the horizon [38], New Scientist article (September 4, 2004)
- Bigelow's Gamble -- Inside the Bigelow Inflatable-Module Plant [39] *Aviation Week & Space Technology* article (September 26, 2004)
- The Five-Billion-Star Hotel [40] Popular Science article (March 2005)
- Low-Earth Orbit, and Beyond! [41] Popular Science article (March 2005), showing plans for moon cruisers and space yachts
- Progress Made on Inflatable Private Space Module [42], space.com article (March 8, 2005)
- Russians Delay First Test Flight of Space Hotel, msnbc.com article [43] (June 7, 2006)
- EXCLUSIVE: Bigelow Orbital Module Launched into Space [44], space.com article (July 12, 2006)
- Bigelow Aerospace Continues Relationship with NASA-JSC for Space Habitat Technology and Private Sector Space Development [45]
- A History of the Genesis I Private Space Module [46] (Space Pragmatism, July 27, 2006)
- Money Backing the Private Space Industry... Part 3--Robert Bigelow [47] (The Space Monitor, February 11, 2007)
- Inflatable space stations take off [48], *NewScientist.com* [49], simulation and Bigelow-released photographs of Genesis I and Genesis II.

References

[1] Sigurd De Keyser (2006-06-04). "Bigelow Aerospace; Russian Dnepr rocket to launch a 1/3-scale Genesis model" (http://www.spacefellowship.com/News/?p=1596). *SpaceFellowship News*. . Retrieved 2010-03-04.

[2] http://www.bigelowaerospace.com

[3] Bigelow's Gamble (http://www.spaceflightnow.com/news/n0409/27bigelow/) *Aviation Week & Space Technology*, 2004-09-26, reprint (authorized), accessed 2010-02-17.

[4] The Americans may still go to the moon before the Chinese (http://www.economist.com/science-technology/displaystory.cfm?story_id=15543675), *The Economist*, 2010-02-18, accessed 2010-03-04.

[5] Mr. B's Big Plan (http://www.airspacemag.com/space-exploration/16047742.html), Geoffrey Little, *Air & Space Magazine*, 2008-01-01, accessed 2010-02-18

[6] NASA turned on by blow-up space stations (http://www.newscientist.com/article/dn18607-nasa-turned-on-by-blowup-space-stations.html), Paul Marks, *NewScientist*, 2010-03-03, accessed 2010-03-03.

[7] "Out There - Genesis I" (http://www.bigelowaerospace.com/out_there/genesis_I.php). *BigelowAerospace.com*. . Retrieved 2007-03-31.

[8] "Genesis II" (http://www.bigelowaerospace.com/genesis_II/). .

[9] David, Leonard (2004-05-24). "Bigelow Aerospace to Tackle Inflatable Space Habitats" (http://www.space.com/news/businessmonday_040524.html). Space.com. .

[10] Fildes, Jonathan (2006-07-14). "Inflatable space module puffs up" (http://news.bbc.co.uk/2/hi/technology/5173388.stm). *BBC News*. .

[11] Lemos, Robert (2007-04-10). "Space Symposium: Bigelow Puts Price on Space -- $15 Million" (http://blog.wired.com/wiredscience/2007/04/space_symposium.html). *Wired Science blog*. . Retrieved 2007-04-11.

[12] . http://www.hobbyspace.com/nucleus/index.php?itemid=18366.

[13] Private Moon Bases a Hot Idea for Space Pioneer (http://www.space.com/businesstechnology/private-moon-bases-bigelow-aerospace-100414.html), Leonard David, *Space.com*, 14 April 2010, accessed 2010-04-15.
[14] The Five-Billion-Star Hotel, 1 Mar 2005 (http://www.popsci.com/scitech/article/2005-03/five-billion-star-hotel)
[15] " CSS Skywalker (http://www.astronautix.com/craft/cssalker.htm)". Encyclopedia Astronautica.
[16] Space Hotel Visionary Proposes Modified "Orion Lite" Spaceship for NASA, 14 Aug 2009 (http://www.popsci.com/military-aviation-amp-space/article/2009-08/space-hotel-visionary-proposes-modified-spaceship-nasa)
[17] Company sees future in space (http://www.lvrj.com/news/9169327.html), *Las Vegas Review-Journal*, 2008-08-15, accessed 2010-05-03
[18] Bigelow Aerospace (2007-08-13). "Special Announcement From Robert T. Bigelow" (http://www.bigelowaerospace.com/news/?Special_Announcement). Press release. . Retrieved 2010-05-03. "This dramatic rise in launch costs has forced us to rethink our strategy with Galaxy. Due to the fact that a high percentage of the systems Galaxy was meant to test can be effectively validated on a terrestrial basis, the technical value of launching the spacecraft - particularly after the successful launch of both Genesis I and II - is somewhat marginal. Therefore, we have decided to expedite our schedule yet again, and are now planning to move ahead directly with Bigelow Aerospace's first human habitable spacecraft, the Sundancer."
[19] "SpaceX Launch Manifest, copyright 2010, 'last update' date unknown" (http://www.spacex.com/launch_manifest.php). *SpaceX.com*. 2010. . Retrieved 2010-02-13.
[20] Bigelow Aerospace (2008-02-05). "Bigelow Aerospace and Lockheed Martin Converging on Terms for Launch Services" (http://www.bigelowaerospace.com/news). Press release. . Retrieved 2008-03-09.
[21] No major hurdles to upgrade Atlas V rockets for people (http://www.newscientist.com/article/dn13290-no-major-hurdles-to-upgrade-atlas-v-rockets-for-people.html), David Shiga, *NewScientist.com* (http://www.NewScientist.com), 2008-02-07, accessed 2010-03-01
[22] David, Leonard (2006-07-12). "Bigelow Orbital Module Launched into Space" (http://space.com/missionlaunches/060712_genesis-1_launch.html). *Space.com*. . Retrieved 2007-03-11.
[23] Malik, Tariq; and Leonard David (2007-06-28). "Bigelow's Second Orbital Module Launches Into Space" (http://www.space.com/missionlaunches/070628_genesis2_update.html). *Space.com*. . Retrieved 2007-06-29.
[24] "Launch of Genesis I Pathfinder Ushers in a New Era of Commercial Space Development" (http://www.spacefellowship.com/News/?p=1616). *SpaceFellowship.com*. 2006-07-15. . Retrieved 2007-03-12.
[25] http://www.bigelowaerospace.com/genesis_II/
[26] Ingham, Jay (2007-02-13). "Genesis I Vehicle Performance Update" (http://www.bigelowaerospace.com/out_there/genesis_I_vehicle_performance.php). *BigelowAerospace.com*. . Retrieved 2007-03-11.
[27] Ferster, Warren (2006-09-21). "Private Space Habitat Could Launch by 2010" (http://www.space.com/news/060921_bigelow_plans.html). *Space.com*. . Retrieved 2007-07-28.
[28] Covault, Craig (2007-04-08). "Bigelow Reveals Business Plan" (http://www.aviationweek.com/aw/generic/story_generic.jsp?channel=awst&id=news/aw040907p2.xml). *Aviation Week*. . Retrieved 2007-07-28.
[29] "SpaceX Launch Manifest" (http://www.spacex.com/launch_manifest.php). *SpaceX.com*. 2009. . Retrieved 2009-09-07.
[30] Klamper, Amy (2009-08-14). "Company pitches 'lite' spaceship to NASA : Stripped down version of NASA's future spaceship could be ready 2013" (http://www.msnbc.msn.com/id/32418057/ns/technology_and_science-space/). Msnbc.com. . Retrieved 2009-09-07.
[31] Clark Lindsey. "Bigelow Aerospace proposes "Orion Lite" concept to Augustine panel" (http://www.hobbyspace.com/nucleus/index.php?itemid=14480). Space Transportion News. .
[32] Private Spaceflight Goes Public (http://cosmiclog.msnbc.msn.com/archive/2010/02/01/2191461.aspx), Alan Boyle, *MSNBC Cosmic Log*, 2010-02-01.
[33] ClarkeFoundation.org (2006-09-12). "2006 Arthur C. Clarke Awards to honor Walter Cronkite and Robert Bigelow" (http://www.clarkefoundation.org/news/091206.php). Press release. . Retrieved 2007-03-11.
[34] SpaceFoundation.org (2007-01-25). "Space Foundation honors Bigelow Aerospace with Space Achievement Award" (http://news.spacefoundation.org/index.cfm?releaseid=5B35E838-FD4B-547F-C6381435F9F1991A). Press release. . Retrieved 2007-03-11.
[35] http://www.thespacereview.com/article/187/1
[36] http://www.spacefellowship.com/News/index.php?s=Bigelow+Aerospace
[37] http://www.reuters.com/printerFriendlyPopup.jhtml?type=scienceNews&storyID=5904247
[38] http://www.newscientist.com/news/news.jsp?id=ns99996347
[39] http://www.spaceflightnow.com/news/n0409/27bigelow/
[40] http://www.popsci.com/popsci/aviationspace/e08989c49db84010vgnvcm1000004eecbccdrcrd.html
[41] http://www.popsci.com/popsci/aviation/article/0,20967,1027555,00.html
[42] http://www.space.com/missionlaunches/050308_bigelow_update.html
[43] http://www.msnbc.msn.com/id/13171475
[44] http://space.com/missionlaunches/060712_genesis-1_launch.html
[45] http://technology.jsc.nasa.gov/bigelow_story.cfm
[46] http://spacepragmatism.blogspot.com/2006/07/history-of-genesis-i-private-space.html
[47] http://spacemonitor.blogspot.com/2007/02/money-backing-private-space-industry.html
[48] http://www.youtube.com/watch?v=Ai6_8GK-R0I
[49] http://www.NewScientist.com

Space Island Group

Space Island Group (**SIG**) is a commercial organization based in West Covina, CA that is dedicated to the development of commerce, research, manufacturing and tourism in space. They plan to accomplish this by designing, building and operating commercial space transportation systems and destinations. Their flagship project is the Space Island Project. This will be a stand-alone commercial space infrastructure supporting manned business activities in Low Earth Orbit (LEO) 400 to 500 miles above earth. They plan to accomplish this through the use of technologies, vehicles and procedures developed by NASA and aerospace companies over the last 25 years.

Funding

One of the unique propositions outlined by the Space Island Group is their funding methodology. They argue that previous attempts at commercial space projects have had a dependency on revenue and subsidies from government agencies[1], but that they have a funding plan that would be viable through capitalizing on commercial markets such as solar power satellites.

External links

- The Space Island Group web site [2]
- The Space Island Group solar power satellite initiative [3]

References

[1] http://www.spaceislandgroup.com/nasacost.html
[2] http://www.spaceislandgroup.com
[3] http://www.spaceislandgroup.com/solarsat.html

Space Island Project

The **Space Island Project** is a project whose goal is to create a stand-alone commercial space infrastructure supporting manned business activities in Low Earth orbit (LEO) 400 to 500 miles above earth.

The centerpiece of this project is the Space Solar Energy Initiative which will collect energy with solar panel satellites and then beam this energy to receiving towers on Earth using "very weak microwave beams." [1]

The Space Island Group is the organization behind this project, and plans to accomplish this through the use of technologies, vehicles and procedures developed by NASA and aerospace companies over the last 25 years.

See also

- Commercial space hotels

External links

- The Space Island Group web site [2]
- The Space Island Project overview [2]

References

[1] http://www.spaceislandgroup.com/solarspace.html
[2] http://www.spaceislandgroup.com/programs-services.html

Space Tourism Society

The **Space Tourism Society** (**STS**) was founded in 1996. It is the first organization specifically focused on the space tourism industry.

Their stated goals, quoted from their website [1]:

> "To conduct the research,
> build public desire,
> and acquire the financial and political power
> to make space tourism available
> to as many people as possible
> as soon as possible."

STS has chapters in Japan, Norway, and the United Kingdom.

The STS is a California 501(c)3 nonprofit organization based in the USA. STS aims to provide the vision and voice for the evolution of humanity off-world in a humane, fun, and beautiful direction. STS was created to inspire people to build real products for future use in space.

See also

- Space tourism
- Space Adventures
- Space colonization
- Private spaceflight
- Commercial Astronaut
- List of private spaceflight companies
- Quasi Universal Intergalactic Denomination

External links

- Space tourism Society [2]

References

[1] http://www.spacetourismsociety.org
[2] http://www.spacetourismsociety.org/

Article Sources and Contributors

Outer space *Source*: http://en.wikipedia.org/w/index.php?oldid=368320584 *Contributors*: 11harj, 3velvet3, A455bcd9, Aaron7chicago, Abc518, AbsolutDan, Academic Challenger, Adumb50, Aeluwas, Ageekgal, Akendall, Akriasas, Aldie, Alexwcovington, Alistair Stevenson, Alkhowarizmi, All Is One, Allen234, Allstarecho, Amplitude101, Andykass, Antandrus, Anxietycello, Aotake, Applehead77, Archangel Lucifer, Ary29, Ashley Y, Ask123, Atlantia, Atulsnischal, Ausdortmund, Balcer, Beefyt, Beetstra, Beezhive, Beland, Bensaccount, BlaiseFEgan, Blibberblabber, Blipadouzi, Bobo192, Bogey97, Bongwarrior, Boris Allen, Bradjamesbrown, Bradtcordeiro, Breakingsilly, Brews ohare, Brianna4262, Butterflymind, CVJX, Calmer Waters, Canthusus, Celarnor, Cenarium, Chaim Shel, CharlesC, Childhoodsend, Chmee2, Chris4682, Christopher Lamothe, Chuckusseb07, Circeus, Civil Engineer III, Ck lostsword, Ckatz, Clarince63, Cockneyite, Coffee, ConradPino, Craigy144, Crazilla, Cyclopia, CyrilB, D, D hanbun, DD Ting, DDima, Dantadd, David Shear, David s graff, Davidprior, Dbiel, Deboerjo, Debresser, Declare, Demmy, Dentedface, DerHexer, DesertAngel, Dinohunter, Dionyseus, Discosebastian, Discospinster, Dmitry Brant, Docu, Dominic Hardstaff, Don't give an Ameriflag, Doppy16405, Drbreznjev, Duk, Duncan, EarthPerson, Edecker1969, Edman274, EdoDodo, Egil, El C, Elbigger1, Ems57fcva, Enormousdude, Epbr123, Eras-mus, Eric119, Esrever, Etafly, Eteq, EurekaLott, Excirial, Explicit, FFLaguna, Faelomx, Fail, Fartherred, Fastily, FayssalF, Femto, Finlay, Footwarrior, Fountains of Bryn Mawr, Fplay, Froth, Frédérick Lacasse, Future Perfect at Sunrise, GW Simulations, Gamer007, Gangsterls, Garthmck, Gcanyon, Gene Nygaard, Gene Poole, Gesiwuj, Giftlite, Gillean666, Glane23, Goldenegeal, Gopal81, Gorman, Gracenotes, Gunnar Hendrich, Gurch, Gwernol, H, H falcon, Hairy Dude, Hatch68, Headbomb, Hellotarget, HenryLi, Herbee, Hiddekel, Hojimachong, Hongooi, Hope(N Forever), HorsePunchKid, Hqb, Hswe, Husond, Huzah, IRP, IanA, Iantresman, IceUnshattered, Insanity Incarnate, Intgr, Iridescent, J.delanoy, JEN9841, JForget, JFreeman, JTN, JaGa, JackSchmidt, Jaffer, Jamie C, Jatkins, Javit, Jbshryne, Jcoppo, Jdrewitt, Jeandré du Toit, Jeffrey Mall, JeremyA, Jh51681, Jigsy, Jkl, Jleon, Jmlk17, Joeardingertaintsociety, JohnnyBGood, José77, Joseph Solis in Australia, JoshuaZ, Journalist, Jrkagan, Jrockley, Jtkiefer, Jugander, Jusdafax, Jyril, Jéské Couriano, KJS77, Ka Faraq Gatri, Kaiba, Karol Langner, Kavanagh, Kazvorpal, Ke4roh, Kingpin13, Kisama, Kitch, Kizor, Kompilator, Kookyunii, Krash, Kukini, Kungfuadam, Kuru, Kyuubi9, Lambiam, Latitude0116, Legis, LeilaniLad, Liftarn, Lightmouse, Lights, Likebox, LilHelpa, Lklovelw1, LonelyMarble, Longhair, Loren.wilton, LuauLarry1969, Lulu71339, MASQUERAID, MER-C, Maheeki, Malcolm Farmer, Mannafredo, Manop, Marauder40, Mark.murphy, Maurreen, Mejor Los Indios, Messhall, Metnik, Michael Hardy, Miguel.mateo, Miraculousrandomness, Mirkoruckels, Mlm42, Mr Pineapples, Mr. Wheely Guy, Mymines229, Mysid, N96, NJDevils44, NYKevin, NawlinWiki, Nbarth, Ndcube, Netsnipe, Neyzen, Nicke L, Nigholith, Ninja Wizard, Njaelkies Lea, NuclearWarfare, Octane, Oda Mari, Onsly, Oo7565, OtakuNOVAkun, OverlordQ, Oxymoron83, PaePae, Pagw, Pahles, Paradoxen, Paxse, Philip Trueman, Pinethicket, Poonie36, Poppafuze, Pred, Pvskirankumar, Quadell, QueenCake, Quiddity, RJHall, RP459, Raining girl, RandomCritic, Ratsbew, Ref2conf, Rich257, Richard Arthur Norton (1958-), Richard001, Rigmahroll, Rjwilmsi, Rmhermen, Robert Brockway, Robert Merkel, Rod57, Ronhjones, Rory096, RoyBoy, Rror, SDC, SElefant, SFGiants, Sabzaby, Sagittarian Milky Way, Samizz, Sanao, Sandahl, Santiparam, Savidan, Sdsds, SeanWillard, Secretmessages, SelfStudyBuddy, Serendipodous, Sesshomaru, Shadowjams, Shaunfin123, Shaunmoss, Shawn in Montreal, Shlomke, Silence, Simetrical, SimonHarvey, Sintaku, Sionus, Slashme, Slavik0329, SnoopY, Someguy1221, Soporaeternus, SpaceFlight89, SpaceGuide, Spacegoat, Spark Moon, Spitfire19, Spliffy, Springnuts, Sprtfshrmn, SpuriousQ, Squeakygoose, Starghost, Statsone, Stephantom, StephenColberto, Steve Quinn, Stevejabs, Stoned Proffesor, SylvieHorse, Symane, Syrian eagle, Szajci, THEN WHO WAS PHONE?, TKD, Taroaldo, Tarotcards, TastyLemonsmmm, Tdadamemd, Terrancommander, Texas™, The High Fin Sperm Whale, The Thing That Should Not Be, The monkeyhate, The-G-Unit-Boss, TheMuuj, Tiddly Tom, Tide rolls, Tiggerjay, Titoxd, Tommy2010, TopGUN71691, TotoBaggins, Trampled, Trusilver, Turkishbob, Twang, Ufviper, UltioUltionis, Ultra megatron, UnitedStatesian, V.V zzzzz, Vahid11, Valerius Myotis, Vanished User 4517, Vegaswikian, Vera Cruz, VinnyR, Violask81976, Vipinhari, Vivio Testarossa, Voltzman, Vranak, Vsst, Vystrix Nexoth, WODUP, Waldir, Wanderingjew, Weeliljimmy, Welshjim, Wikiborg, Wikid77, Wikipedia brown, WildWildBil, Willking1979, Wolfkeeper, Wtmitchell, XJamRastafire, Xanzzibar, Yamamoto Ichiro, Yath, Ybbor, Yidisheryid, YouthoNation, Yt95, Ytny, Ytrottier, Yupik, Zaharous, Zaiken, Zozza, Zurrr, Île flottante, ملاع بوبحم, 731 anonymous edits

Human spaceflight *Source*: http://en.wikipedia.org/w/index.php?oldid=368678873 *Contributors*: (, 216Kleopatra, Abebenjoe, Adam78, Ageekgal, Ahoerstemeier, Ahpook, Aitias, Alex.tan, Alfio, Algocu, Altenmann, Ancheta Wis, Anoop, Antennaman, Anxietycello, Aodin, Apoc2400, ArglebargleIV, ArielGold, Art LaPella, Astudent, Audin, Auric, Australisian, Az29, Beetstra, Bergsten, Betterusername, Binary TSO, Bob rulz, Bryan Derksen, Bubba hotep, BuffaloChip97, By78, Byrial, Calypso, Capzilla, Chafis, Challisrussia, Clerks, Cncs wikipedia, Colds7ream, Colin Douglas Howell, Conversion script, Craigboy, Curtains99, DanSlotea, Daniel Quinlan, Danorton, Darth Panda, David Kernow, David136a, Deathbyday, Denelson83, Dennis Brown, DethmeOw, Dialectric, DocWatson42, Dojarca, Dr U, Dthx1138, Dutchdrummer99, Ed Cormany, Edward, Egg plant, Epbr123, Eric-Wester, Ericmachmer, EurekaLott, Evil Monkey, Factitious, Falcon8765, Fnlayson, Future Perfect at Sunrise, Fuzheado, GCarty, GW Simulations, Gdavidp, Gilliam, GrampaScience, Greyhood, Grioghair, Ground Zero, Gwillhickers, Hardscarf, Harsha363, Heirpixel, Hiberniantears, Iluvcapra, Imroy, Interiot, International space agency, JCDenton2052, JHMM13, JTN, Jaffer, Jagged 85, James xeno, JamesAM, Jamesontai, Jatkins, Jayann, Jeepday, Jeff G., Jennavecia, Jesternaut, Joema, Johnxxx9, Jonathunder, Joseph Solis in Australia, Jrkenti, Jschwa1, Julesd, Kaiba, Kintetsubuffalo, L3p3r, Leoboswell, Leon7, Lexicon, Lightmouse, LilHelpa, Lissajous, Loren36, Luk, M.nelson, MBK004, MYury, Mahjongg, Man on the Mars, Martijn Hoekstra, Materialscientist, Matthead, Mav, Mawa, Mayfare, Mayspaceroxx, Mcorazao, Mellisa Anthony Jones, Michael Hardy, Mike Rosoft, Mo7amedsalim, Monadman, Moonriddengirl, Mountain, Mourn, Murlough23, N2e, Narayan, Nfutvol, Nicholasnice, Nikola Smolenski, NuclearVacuum, Nv8200p, Obakeneko, OllyG, Op. Deo, Patrick, Pedant, Pepperpiggle, Pgk, Phantomsteve, PheonixSong, Philip Trueman, Pirates920, Polylepsis, Popageorgio, Puchiko, RadioBroadcast, Raso mk, Ravichandar84, Reubenbarton, Rich Farmbrough, Ricnun, Rillian, Rizzoj, RjLesch, Rjwilmsi, Rlandmann, Rmhermen, Robert Merkel, RockinRob, Roscoe x, Rowan Moore, Rschu, SECProto, Sagdiyev, Samuel Webster, Sanders muc, Sdsds, Shahin.shn, Shanoman, Shawnc, Slakr, Smiller933, Snapdragonfly, Someone else, SpaceFlight89, Stay by me, Stephenb, Supersexyspacemonkey, Swpb, Synchronism, Tanthalas39, Tdadamemd, Tellinghistory, The Transhumanist, Thedeadlypython, ThomasK, Tianxiaozhang, TimShell, Tttecumseh, Txomin, Tyrol5, UberMan5000, Uncle Scrooge, Valador, Vsmith, WilyD, Witchzilla, Wmahan, Wolfkeeper, Wyss, X201, Yki, Yurik, Zippanova, 308 anonymous edits

Human adaptation to spaceflight *Source*: http://en.wikipedia.org/w/index.php?oldid=367033250 *Contributors*: 2c-b, Alansohn, Amorymeltzer, Andy M. Wang, Ann Stouter, Anthony Appleyard, Art Carlson, Axl, Beno1000, Bobblewik, Bobby D. DS., Brianga, Burschik, CapitalLetterBeginning, CerealBabyMilk, Chris the speller, Chuuumus, Crhayes88, Cyrius, DMacks, DO11.10, DanielCD, DocWatson42, E Wing, EAAA, Eassin, Eatsaq, Enviroboy, FRA, Fangz, Favonian, Figma, Fredrik, Furrykef, Fyrius, Geeoharee, Gilliam, GraemeL, Graham87, Hairy Dude, Helge Skjeveland, Horkana, Hugonius, IanOsgood, Ilikefood, J'raxis, JaGa, Jhf, Jj137, Jjamison, JoeBruno, Jon Monreal, Jrockley, Jschwa1, Julia Rossi, Keilana, Kingdon, Kizor, Knowledge Seeker, Larry V, Max Naylor, Mccready, Midgley, Neparis, Night Gyr, Noclevername, OllyG, Olorin28, Oxymoron83, Pathoschild, Phantom in ca, PoccilScript, Proofreader77, Qz, Rich Farmbrough, Rickterp, Rjwilmsi, Sam Hocevar, Sam8, Sandius, Sdsds, Shiftchange, Sophistifunk, Sstoneb, Suniti karunatillake, SwedishPsycho, Tango, TheAdventMaster, TheFutureAwaits, Tierlieb, Tommyt, Tresiden, UnitedStatesian, VernoWhitney, Viriditas, Woohookitty, Ytrottier, Zepheus, 102 anonymous edits

Space medicine *Source*: http://en.wikipedia.org/w/index.php?oldid=364184909 *Contributors*: Abune, Ageekgal, AndreasJS, Angr, Boromir123, Clicketyclack, D Dinneen, Elonka, Emeraude, Fractain, Intelligentsium, Jef-Infojef, MasterOfHisOwnDomain, Memstone, Phantom in ca, Primalmoon, Rbejoy, Shannon bohle, Spalding, Sstrauss, Sverigekillen, Swpb, Tanthalas39, True Pagan Warrior, WhatamIdoing, Wolfkeeper, Xerxes247, 21 anonymous edits

Space adaptation syndrome *Source*: http://en.wikipedia.org/w/index.php?oldid=368339835 *Contributors*: A Doon, Ageekgal, Alexander Iwaschkin, Amikake3, Astro jpc, CDN99, Charles Matthews, DanB, Darklilac, Embhee, Enfenion, Evil Monkey, Ex nihil, Gnoitall, Grafen, Jfdwolff, Karada, Kasperl, KnightLago, Kotra, Leafnode, Lee6597, Liberatus, MacRusgail, Mendaliv, Midnightblaze, N2e, NerdyNSK, Noclevername, Ponder, RJHall, Radagast83, Rhombus, Rillian, Shimgray, Simulation12, Skizzik, Swpb, Tdadamemd, Vardion, 16 anonymous edits

Space tourism *Source*: http://en.wikipedia.org/w/index.php?oldid=368511631 *Contributors*: 1bevingtonco, Aagg123, Aaronwinborn, Adam78, Aeural, Ageekgal, Agmon, Ahoerstemeier, Akradecki, Alansohn, AlexiusHoratius, Allanlw, Anarchist42, AngelOfSadness, Angela, Anxietycello, Argo Navis, ArielGold, Armigo, Arnero, Astro jpc, Astudent, Avinesh, AxG, B. Fairbairn, Babyrina2, Battamer, Beardo, Benbest, Benc, Betterusername, Bhludzin, Biawaknaga, BillCook, Binba, Bobblewik, Bongle, Bongwarrior, Booksworm, Bovineone, Bradjamesbrown, Briaboru, Btanders, Butros, CaptainAmerica, Centurion328, Chad Vander Veen, Chanlyn, Claymorgan, Cmdrjameson, ColdCase, Cometstyles, Cool Blue, Crypticfirefly, Cs302b, Ctbolt, Cwastell, Czerka Corp., Dahveed323, Dan Guan, David Latapie, Daviiid, DeadEyeArrow, Delldot, Dev1n, Dgmoran, Dhaluza, Dhartung, Dispenser, Dojarca, Dputig07, Dragonbeast, DuncanHill, E0N, EarthPerson, Edward, Egarolera, El gregor, Elatanatari, Ellol, Emperorbma, Enormousdude, Eric-Wester, Esn, Espacetickets, Executor Tassadar, Fahrenheit451, Felix Wan, Francis45, Froth, Fsotrain09, Fuhghettaboutit, G Clark, Gaius Cornelius, Gangsterls, Gentle Reader, Gerrit, Gleam, Glloq, Gogo Dodo, Graue, Greatestrowerever, GregorB, Greyhood, Grishalevit, Ground, Gtstricky, Gwernol, GwydionM, Hack-Man, Hateless, Hede2000, Hektor, Hello45044, HieuNV, Hu, Hu12, Hux, Hydrokevin, IMC.esq, IanOsgood, IddoGenuth, Ilvar, IncognitoErgoSum, Infophile, IstvanWolf, JHuwaldt, JYi, Ja 62, Jack1981, James Doehring, JarlaxleArtemis, Jatkins, Jazmatician, Jessi1989, Jmackaerospace, Joffeloff, John, Johntex, Joncnunn, Joseph Solis in Australia, Joshbrez, Josinj, Jpatokal, Jrockley, Kajmal, Kazrian, Kember, Kernel geek, Killingentelletilti, Kiore, Kirokita, Kmweber, Koselara, Kotra, Kozuch, Kzuyev, Latitude0116, LegitimateAndEvenCompelling, Lihaas, Lilac Soul, Loganis, Lolnostanku, LovesMacs, Lyonslions, MBK004, MER-C, Mac, Maelnuneb, Malcolm Farmer, Man on the Mars, Markchuaklang, Martarius, Mattgenne, Matthew Stannard, Matthias5, Mav, Mazarin07, Mboverload, Mendaliv, Michaelbusch, Miketwo, Miyokan, Modster, Mtminchi08, Mushroom, N2e, N328KF, Nairebis, Namiba, Nasa-verve, NeuronExMachina, Nevit, Nickshanks, Nitishgulati, Noclevername, Nuno Tavares, ObiusX, Obrienz, Ohnoitsjamie, Old Guard, Oleg Alexandrov, Oliphaunt, Omicronpersei8, Onsly, PBP, Pah246, Pahles, Paranoid, Patrick, Pattern86, Paul Weaver, Peyre, Philip Howard, Piano non troppo, Plasticup, Praetonia, Qa Plar, Qwerqwerqwer, RBBrittain, Radon210, Raerth, Rajeevmass, RandomCritic, Rangi42, Raven in Orbit, RedWolf, Rich Farmbrough, Rickremember, Rillian, Rjwilmsi, Rlandmann, Rmhermen, Robert Merkel, Ronnie Nader Bello, Runningman85, Rwxrwxrwx, SCEhardt, Saganaki-, Sanao, Sanchom, SchuminWeb, Scjessey, Scroteau96, Sdsds, SeNeKa, ShimaKatase, SigmaEpsilon, Sims2004, Skybum, Sladen, Solipsist, Sootyboy, Spaceman13, Spute, Squash, Star Mississippi, Stealthound, Steverit, Struway, Supapuerco, Sverdrup, Swatje, Takeel, Taksim25, Tang23, Tango, Tdadamemd, Tellinghistory, ThatBajoranGuy, The Thing That Should Not Be, The not, Thecomputist, Thue, Tony1, Travelwriter222, TreasuryTag, Tsujigiri, Turkeyphant, U5K0, Vald, Vegaswikian, Vishnava, Voyevoda, Wavelength, Wizardimps, Wolfkeeper, Woohookitty, Wrad, Wronkiew, Wyllium, X201, Yakushima, Yonatan, Yono, Zafiroblue05, Zentropa, Zero1328, کشرز, 476 anonymous edits

Spaceflight participant *Source*: http://en.wikipedia.org/w/index.php?oldid=368096210 *Contributors*: ArielGold, Bcjordan, CommonsDelinker, D6, Florentino floro, GW Simulations, Hektor, JaterGirl, Jatkins, Josinj, MBK004, Neutrality, Ohms law, PsyberS, Radagast, RandomCritic, Rillian, Sdsds, SeNeKa, Sladen, SpaceHistory101, TheDJ, Van helsing, Wongm, ZK001, 13 anonymous edits

Soyuz (spacecraft) *Source*: http://en.wikipedia.org/w/index.php?oldid=367562894 *Contributors*: (, .:Ajvol:., Aalox, AlainV, Alkivar, Altitech, Anders E Larsson, Andre Engels, Andy120290, AntonBreusov, ArielGold, Ashley Pomeroy, Astromp, Audin, Avmich, Basicer, BerserkerBen, Bletch, Bobblehead, Bobblewik, BrendanSugrue, Bricktop, Bryan Derksen, Bwfrank, C1010, Canterbury Tail, Capnned, Cardinal2, Cassius1213, Chairboy, Chrisdnson, Ckatz, Clh288, Craigboy, Cricketmania, DO'Neil, Dan100, DaveHolland, Debivort, Didero, Dimadick, Douglasfrankfort, Dwiki, Ely1, Enceladus, Enigmatzu, Error 404, Europrobe, FairoozAdams, Fetofs, Filanca, Foobaz, GT009No1, GW Simulations, Gadren, Gits (Neo), GregLoutsenko, Halibut Thyme, Hayabusa future, Hede2000, Hektor, Hellisp, Henristosch, Hippietrail, Hobbes78, HordeFTL, Hu12, I12681, Ian Rose, Ianb, IsaacW1982, JTN, Jkeiser, Johnxxx9, Jorunn, KFP, Krinsky, Kwamikagami, La goutte de pluie, Lavenderbunny, Le Hibou, LeonardoRob0t, Lightmouse, LinuxPickle, Loren36, MBK004, MaBoehm, MartinPackerIBM, Mcmustang, Mertozoro, Mhollida, Mlm42, NERV, Namcaps, Necessary Evil, Nickst, Nikai, Niteowlneils, Noclevername, Nuaetius, NuclearWarfare, Odie5533, Paat, Parkjunwung, Patrick, PeepP, Piano non troppo, Pigsonthewing, Poppy, Potatoswatter, RDBrown, RandomCritic, Redwallfox, Rhythm, Rich Farmbrough, Ricnun, Rillian, Rlandmann, Roadrunner, RoyBoy, Saintamh, Sandstein, Sardanaphalus, SchuminWeb, Sdsds, Sea diver, SeanWillard, Sewebster, SimonP, Sjl0523, Slipstream, Smihael, Smobserver, Sobolewski, Soyuz113, Spitwater, Stephen Gilbert, Stuyoung38, Tdadamemd, Tex23, Themanwithoutapast, ThreeBlindMice, ThreeE, Thumperward, Tonik, Tony1, Treff89, UrsusArctosL71, Vedran8080, Vegaswikian, Vmenkov, Vonschlesien, Whitepaw, Will Pittenger, Wronkiew, Xanderer, Xyzzyva, Yossarian, ZeroJanvier, 106 anonymous edits

International Space Station *Source*: http://en.wikipedia.org/w/index.php?oldid=368723840 *Contributors*: !Darkfire!6'28'14, (, -=PhotoN=-, 0.39, 345Kai, 61.9.128.xxx, A Human On Earth, A.Cython, A.T.M.Schipperijn, A8UDI, APT, ASOTMKX, Aafm, Aalfa080, Aalox, Aaron Einstein, Achmelvic, ActofF, Addshore, AdjustShift, AdultSwim, Aero1, AgCam, Ageekgal, Agentaway, Ahoerstemeier, AikiHawkeye, Aisforabstract, Aitias, Akirn, Akradecki, Akut, Alandavidson, Alansohn, Alba, Alerante, Alex Douglas, Alex the dude, Alex.tan, AlexiusHoratius, Alexjohnc3, Alexwcovington, Aliazimi, Alikaalex, Alinor, All Is One, Allstarecho, Alois, Alsandro, Altitech, AmbassadorShras, Ameliadaisy, Amm582, AnOddName, AndersFeder, AndersL, AndreMA, Andrea105, Andres, Andrewpmk, Andy120290, AndyLiNY, Andylkl, Animum, Anonym1ty, Anonymous6494, Antandrus, Anty, Apalapala, Apoc2400, ApothThaeos, Arakunem, Aram33, Archelon, ArdClose, Aremisasling, Aridif, ArielGold, Armeria, Arpingstone, Arsonal, Art LaPella, Assassin3577, Astronaut, Astrowikizhang, Astudent, Atrian, Audriusa, AussieLegend, Avatarcourt, Avillia, Awatral, Awickert, Azazyel, BBnet3000, BRUTE, Ball of pain, Bark, Bastique, BatteryIncluded, Bcrowell, Bdesham, Bdjake57, Bearsalley34, Ben-Zin, Benbest, Bender235, Bennetto, Berry2K, Bevo, BigEdit, BiggKwell, Bigtop, Bill W Ca, BillJohnson0003, Bkell, BlastOButter42, Blockhead9345, BlueMoonlet, Bnjacobs, Bobblehead, Bobblewik, Bobo192, Bodyn, Bonchygeez, Boris Barowski, Boromir123, BowChickaNeowNeow, Bowlhover, Bpqlr, Bradjamesbrown, Brahmanknight, Brainbark, Bricktop, Brj3937, Bryan Derksen, Bubba73, BuffaloChip97, Bugtrio, Bumbi24, Bunchofgrapes, Butros, Butterwise, Buttle, Bwfrank, Byeee, CBDunkerson, CaNNoNFoDDa, Callnishant, Can't sleep, clown will eat me, Canada Jack, Candirocks895, CannotResolveSymbol, Capricorn42, Captian45633, Captmjc, Carbon-16, CardinalDan, Carl w, Carter, CatherineMunro, Cayers2014, Cburnett, Cdogsimmons, CesarB, Chcknwnm, Chesnok, Chilin, Chocoforfriends, Chris Roy, Chrisjj, Chrislk02, Christian List, Chuggsymalone, Cinnamon42, Citicat, Cjosefy, Cjthellama, Cjwright79, Ckatz, Ckruetze, Clh288, ClickRick, Closedmouth, Cmoibenlepro, Coat of Arms, Codepro, ColdCase, Colds7ream, CommonsDelinker, ComradeWolf, Confuzion, Conscious, Conversion script, CoolGuy, Coolhawks88, Coreyjune12, Cornforth, Corvus cornix, Cosmic Latte, Cp111, Cpom, Craigboy, Crowstar, Cryonic07, Cs302b, Cst17, Curps, Cyclonenim, Czolgolz, D, DMacks, DSRH, DVD R W, Da monster under your bed, Da-rb, Dabomb87, DailyWikiHelp, Dalen talas, Dan Gluck, Danielpalen345, DarkHumor, DarthVader, DaveHinz, Davenbelle, Davidthegreen, Dawnseeker2000, Daysleeper47, Db099221, DeadEyeArrow, DeadlyAssassin, Debresser, Defender of torch, Delta107, Demmy, Denelson83, DerHexer, Derlay, Dforest, Dick Tits, Diderot, Discospinster, Dlohcierekim, Dodonov, Dojarca, Dontdoit, Dori, Dougofborg, Douken, Downtownblue, Dozols, Dpaajones, Dravecky, Dreamafter, DreamingGod, Drich42, Dricherby, Drivewiki, Drostie, Dskluz, Dsyzdek, Dthierry, DubaiTerminator, Duccio, Dudegalea, Duffman, Duncan.france, Dvavasour, Ealdgyth, EamonnPKeane, EarthPerson, Earthworm Makarov, EasyTarget, Ecoman24, Ed g2s, EdBever, Ekalb, Electron9, Eleos, Emerson7, Emilfaro, EmirA, Energy Dome, Enviroboy, Epbr123, Eptin, Eragon is the lord of the rings, ErgoSum88, Eric-Wester, Ericdn, Error 404, ErsatzHatrack, Esprit15d, Etr52, Ettrig, Eubulides, Euchiasmus, Everyking, Evil Merlin, Evil Monkey, Excirial, Exir Kamalabadi, Extra999, Extraordinary, Falcon 1986, False vacuum, Famine, Farhan-Tre, Fcn, Feedmecereal, Felinoel, Felyza, Fern 24, Fetchmaster, Fethers, Fhimpe, Fieldday-sunday, Fifth Rider, Fireswordfight, FirstPrinciples, Flamejob, Flewis, Flybywire e2c, Fool4jesus, Fraantik, Fractain, Fragglet, FrankTobia, Frecklefoot, Fred Bradstadt, Fred Stober, Freddicus, Friendly person, Fubar Obfusco, Fukumoto, Full Shunyata, Fullstop, Fumo7887, Funkyflames08, Futbol180, GDonato, GSV Ethics Gradient, GW Simulations, Gabriel, Gaius Cornelius, Gary King, Gazimoff, Gbeeker, Gcm, Gdavidp, Gekritzl, Gene Nygaard, Gentgeen, Geoffrey.landis, Geremia, Gerrit, Ghalas, Giftlite, Gimboid13, Giradvance, Glen Morris, GluonBall, GoRight, Gogo Dodo, Goingareas, Good kitty, Graham87, GrahamColm, GrandDrake, Greg Kuperberg, Gridejp, Ground Zero, Gsaup, Guigo, Gunter, Gurch, H0n0r, HAL11000, Hack-Man, Hamiltondaniel, Hamiltonstone, Hanskarlperez, Hanweipaulchen, Hanzoku, Harp24, HarryAlffa, Hazydan, Hdt83, Hdw, Hektor, Heman5up, Heron, Hi.ro, HiLo48, Hibernian, HumphreyW, Huntster, Hut 8.5, Hydrargyrum, Hydroksyde, Hymyly, IanOsgood, Icairns, IceDragon64, Idiotic ally, Ijc1982, Ikanreed, Ikh, Ikyork, Il hamster, Imsaguy, Indosauros, Infrogmation, Ingolfson, InternationalGayStation, Intothewoods29, Irdepesca572, Iridescent, Is he back?, Itai, J.delanoy, J929, JBDRanger, JEckart, JForget, JJJust, JLogan, JNW, JRWalko, JaGa, Jachim, JackSchmidt, Jackfork, Jagged, Jake Wartenberg, James Doehring, Jamesday, Jamie2710, Jaques O. Carvalho, Jason C.K., JasonAQuest, Jatkins, Jclemens, Jeanenawhitney, Jedi Shadow, Jeff G., Jeffq, Jer10 95, Jfiling, Jfisherhr, Jhaldenwang, Jiddisch, Jimp, Jkl, Jmathies, Jmfj, Joaopais, Joe 1987, John, John Boone, John Doe or Jane Doe, John Reaves, John oh, JohnC, Johndrinkwater, Johnwcowan, Joke137, Jordan.Kreiger, JorgeGG, Joseph Solis in Australia, Jossi, Jpk, Jpvinall, Jtkiefer, Jumping cheese, Juxi, Jwrosenzweig, KSC-on-orbit, KYN, Kaaveh Ahangar, Karanacs, Karn, Kavanagh, Kaztec, KeithCu, KeltRanger, Kember, Kendrick7, Kerrio, KevM, Kgasso, KnightLago, Kolt, Korpios, Kotra, Koyaanis Qatsi, Kozuch, Ktr101, Kudret abi, Kukanotas, Kyle 314159, Kyle Barbour, Larryzie, Leafyplant, Lear's Fool, Lee, Lee Pavelich, LeeHas, Lekoman, Leon7, LeonardoRob0t, Lestatdelc, Leuko, Lighted Match, Lightmouse, Lilac Soul, Lino Mastrodomenico, Linuxbeak, Little Mountain 5, Lomedae, Looxix, LorenzoB, LostLucidity, Lot49a, Lotse, Lsinrc, Luk, Lulu71339, Lumos3, Lycurgus, MARussellPESE, MBK-iPhone, MBK004, MLilburne, MSFT, Madchester, Madhero88, Malderi, Malleus Fatuorum, Man on the Mars, ManBearPig69, Manormadman, MarcS, MarkSutton, MarkV, Markpeak, Marsbound2024, Martarius, Martin451, MartinHarper, MartinPackerIBM, Materialscientist, Matevzk, Matiasmoreno, Mav, McSly, Mcpx, Mdf, Mek20, Melonhead, Memming, Mephiles602, Mercury, Mhsb, Michael Devore, Micromaster, Miguel232, Mike Dillon, Mike Peel, Mike6271, Mikeeg555, Mini-Geek, Mion, MisfitToys, Misza13, MithrandirAgain, Mithridates, Mlm42, Mlnovaaa, Mmeijeri, Mnw2000, Mo0, MoRsE, Modest Genius, Modster, Modulatum, Montymark, Moofpocket, Moogin, Moondyne, Morinohtar, Mr Stephen, Mreay, Mrh30, Mrwojo, Ms2ger, Mschel, Mschribr, Mt2131, Mustang dvs, Mxn, Mysid, N2e, N419BH, NGX463, Naelphin, Nakon, Natl1, Navy blue84, NawlinWiki, Nborders1972, Necessary Evil, NellieBly, Nerrolken, Network57, NeuronExMachina, Neutrality, NicholasSThompson, Nickj, Nickshanks, Nidator, Nihilismmm, Nike'97, Ninetyone, Ninly, Nisse321, Nitya Dharma, Nixer, Nk, Nmebrown, Nneonneo, Nodi25, Noha307, Norlanson, Nrpf22pr, Nuaetius, Numbo3, Nuno Tavares, Oblivious, Offliner, Ohconfucius, Ohnjaynb, Okihita, Okivekas, Oleg Alexandrov, Olegwiki, Oliphaunt, Oliverdl, Olivier, Olivier200, OlkhichaAppa, Omgee, Omicronpersei8, Omnieiunium, Onceonthisisland, Oneiros, Onsly, Opk, Originalwana, Oscarthecat, Osomec, Osssua, Otterz Rok, Pakaran, PaladinWhite, Papa November, Paranoid, Patrick, PatrikR, Pax85, Pbroks13, PeepP, Pegasus1138, Penubag, Peripitus, Peteness12, Peter, Peter Ellis, Peter L, Peyre, Phansen, Pharos, PhilKnight, Philip Howard, Philopedia, Phoenix2, Pie Man 360, Pie4all88, Pigby, Pigsonthewing, Pingveno, Pinktulip, Pishogue, PistolPete037, Pixel8, Pjrich, Plasticup, Plautus satire, Populus, Porqin, Possum, Potatoswatter, Preetikapoor0, Prolog, Proofreader77, Pumpmeup, Punarbhava, Punk boi1221, Q43, Quadraxis, QueenofBattle, Quijote3000, Qxz, RB McLeroy, RJHall, RK, RacingThree333, Radagast, RadioFan, Radtek67, Ragesoss, Ran4, RandomCritic, Rangek, Ranger Steve, Rattlesnake, Raul654, Rbs, RealGrouchy, RedBLACKandBURN, Reedy, Remember the dot, Remotelysensed, Renesis, RenniePet, Res2216firestar, Resplendent, Reubenbarton, RexNL, Reywas92, Rich Farmbrough, Richhaddon, Ricnun, Rillian, Rjanag, Rjwilmsi, Rlevse, Rlpreece, Rmhermen, Rnt20, RoachMcKrackin, Roadrunner, RobNS, Robert K S, Robert Merkel, Robert Skyhawk, RobertG, Robzz, Rocketmaniac, Rod57, Rokfaith, Ronark, Ronhjones, Ronsmytheiii, Ropers, Rossenglish, Rotem Dan, Roxi2, RoyBoy, Rukaribe, Ruslik0, RxS, Ryanmac06, Røed, S142968, SGGH, ST47, Sagittarian Milky Way, Salmar, Salsa man, Salsburyway, Sam Clark, Sam8, SandyGeorgia, Satori Son, Savidan, SchuminWeb, Scoobing, Scotthatton, Scotthodgins, Sdornan, Sdsds, SeNeKa, Sean William, SeanMack, SeanWillard, Search4Lancer, Sebcastle, Seifip, Serknap, Seth Ilys, Sfahey, ShakataGaNai, Shanes, Sharkrocket, SheldonKalnitsky, ShelfSkewed, Shibo77, ShimaKatase, Shimgray, Shoemakerleve9, ShuttleDiscovery, Sinigagl, Sjakkalle, Sjones23, Skela, Skinnyweed, Skizzik, SkyWalker, Slammer111, Sleigh, Sligocki, Smartech, Smcgray303, Smolarek, Sniperz11, Snowflake382, Snowolf, Soulkeeper, Southern Illinois SKYWARN, Spaceaholic, Spaceman13, Spacestationinfo, Speedyj, Spiel496, Splette, Srain, Sry85, Stanley1100, Starks, Starnestommy, StaticGull, Statsone, Stephen G. Brown, Stizz, Stmoose, Strich3d, SuperDude05, SuperTyphoon, Superbuffman, Superm401, Supersexyspacemonkey, Superzohar, Supspirit, Swalker20659, Swid, Symane, Synchronism, Szumyk, Tagishsimon, Tariqabjotu, Tarquin, Tassedethe, TastyCakes, Tcwd, Tdadamemd, TeH nOmInAtOr, Techman224, Tegla, TehBrandon, Tellinghistory, TerrenceandPhillip, The Anome, The Singing Badger, The Thing That Should Not Be, The ed17, The posp, The wub, TheDJ, TheDmanator1000, TheSpellchecker, Theflyer, Thejerm, Thelem, Themanwithoutapast, Theoneintraining, Thierryc, Thomas Connor, ThorX13, ThreeE, Thunderbird2, Tiddly Tom, Tide rolls, Tim Keighley, Tim1357, Timc, Tiptoety, Tkeator, Tlc61, Tmotom, Tom walker, Tommy2010, TommyMoullet, Tompagenet, Tomsmith0002, Tomwalden, Tony1, Tonywalton, Topher0128, Torritorri, Tough Little Ship, Towne, Toxicroak, Tpbradbury, TraxPlayer, Treenaks, Trevgreg, Trevyn, Truthkeeper88, Tsunomaru, TungstenCarbide IV, Tuvas, TyeCowan, Tyrol5, U5K0, Uhrtsgdsf, Ultraexactzz, Underwoodl06, Unixxx, Unschool, Unyoyega, Urmom515, UsagiYojimbo, User F203, VMS Mosaic, Ve3wtj, Vegaswikian, Vgranucci, Vildricianus, Vincentz, Violentbob, Voidxor, Vsst, Walter, Warofdreams, Wavelength, Waynerifish, Weatherman1126, Weeliljimmy, Welsh, WereSpielChequers, Wereon, Wernher, Wetman, WhatisFeelings?, Whoop whoop, Whostolemypen, Wiki8fishy, Wikipedia991, Wikipedical, Wildhartlivie, WinterSpw, Wirehead, Wnt, WojPob, Wolf530, Wolfkeeper, WolfmanSF, WriterHound, Wronkiew, Wtmitchell, Wwheaton, XJamRastafire, Yanksox, Ysangkok, Zundark, Zyrkon, Ævar Arnfjörð Bjarmason, 朝彦, 1842 anonymous edits

Dennis Tito *Source*: http://en.wikipedia.org/w/index.php?oldid=365067697 *Contributors*: Ageekgal, Aron Håkanson, Blindmanonthemountain, Bongwarrior, Booyabazooka, Brandon97, Bricktop, CanadianLinuxUser, Cnull, CommonsDelinker, Curps, D6, DannyDaWriter, Danski14, Davidcannon, Denelson83, Dewet, Elendil's Heir, Enchanter, Engology, Espace, Fredrik, GeeJo, Gits (Neo), GregorB, Guat6, Hektor, Histrydude, Hux, Immunize, Infrogmation, Jacek Kendysz, Jaraalbe, Jatkins, JayRH, Jeltz, Kember, Kotra, Lord Hawk, Lupin, Lurker, Madhero88, Marsbound2024, Mordamir, Mufka, N328KF, Nikai, Nixer, Nobunaga24, PBP, Paranoid, Pbleic, Peteharding, Psychotroll, Rajeevmass, Reinoutr, Rillian, Rje, Salsa Shark, Sanao, SchuminWeb, Shanes, Shoeofdeath, Snori, Sootyboy, Square87, Stefanomione, Sveter, TheRingess, Tianxiaozhang, Tide rolls, Tyrol5, Urbanrenewal, Voyevoda, Vroman, Waffle77, WhisperToMe, 68 anonymous edits

Mark Shuttleworth *Source*: http://en.wikipedia.org/w/index.php?oldid=366786047 *Contributors*: 16@r, 555, Aaronbrick, Ageekgal, Ahoerstemeier, Ahunt, Aljullu, Allenapnivag, Aomarks, ArielGold, Arthurbuliva, Atrian, Attys, Bender235, Benjamin Mako Hill, Bletch, Burgundavia, BurnDownBabylon, Bwithh, CambridgeBayWeather, Cazo3788, Chealer, Chic happens, Ciphergoth, Commander, CommonsDelinker, Conti, Cosmonaut27, Crazyboy899, D6, Darklock, Darth Panda, Date delinker, Dawnseeker2000, Dbenbenn, Ddxc, Deleteme42, Dewet, Diwann, Dsym, Edwulf, Elf-friend, Emc2, Energizing mouth spray, Ergil, Escape Orbit, Espace, Evercat, Fanatix, Farm, Fleminra, Fluteflute, Frap, Gaius Cornelius, Gavri, Gecg, Gehirn, Ggenellina, Ghepeu, Gilesmorant, Goat77, Good Olfactory, Guyjohnston, Hackmeister, Hairy Dude, Halaqah, Happenstance, Hektor, Hhielscher, Histrydude, Hoot, Htaccess, Hux, Iluvcapra, Immunize, Impi, Indexum, J.delanoy, JC510, Jaderaid, Jahilia, Jakub, Jaymacdonald, Jcw69, Jeltz, John, Joy, Kablammo, Kacembepower, Kazak, Keilana, Keithkml, Kesla, Kilps, Kishoreasok, Kozuch, Laddiebuck, MaXiMiUS, Mahanga, Man with two legs, Markpeak, Masaki K, Matt Crypto, Mayumashu, Medovina, MikeCapone, Mushroom, N328KF, Nano, NerdyNSK, Nickshanks, Niels Brons, Nikai, Nimakha, Nitot, NoDepositNoReturn, Noisy, Nomad, Ohwell32, Oriez, Oska, Owenozier, PTPP, Pictureuploader, Plasticup, Quarl, R.O.C, Rahuls, Reelrt, Reinoutr, ReyBrujo,

Richhaddon, Rillian, Robert25, SF007, SchuminWeb, Secretlondon, Shado.za, Sigmundur, Simxp, Sladen, Snori, Socrates2008, SodiumHydroxide, SqueakBox, Storkk, Sveter, Technobadger, Thayvian, The bellman, Thirdmonkey, Tiucsib, Tony Sidaway, Tyrol5, Vaan0910, Varano, Viajero, Vincentt, Wikiboozi, Wizzy, WölffReik, Yaco, Yworo, Ævar Arnfjörð Bjarmason, 191 anonymous edits

Gregory Olsen *Source*: http://en.wikipedia.org/w/index.php?oldid=368548610 *Contributors*: A. Falcao, Alansohn, Argo Navis, Boleyn, Bricktop, Bucs10, Bulgaroctonus, Caerwine, Cantus, Chanlyn, CharlesC, Conti, Cosmonaut27, Espace, Essjay, Fxer, GCW50, Gaius Cornelius, GeeJo, Hektor, Hux, Im.sarcastic, IstvanWolf, JYOuyang, Jackqu7, Josephglantz, Kember, Li-sung, Light current, LilHelpa, Marianocecowski, Matt.whitby, Mipadi, Mithridates, Namiba, Njalle, Nobunaga24, Oravelon, Paul A, Philip Stevens, Pilardi, Reinoutr, Rillian, Rjp0i, Rjwilmsi, Sea diver, Slicky, TheRingess, Tomwalden, Ustuss, Wernher, Wesman83, Zer0faults, 37 anonymous edits

Anousheh Ansari *Source*: http://en.wikipedia.org/w/index.php?oldid=364816291 *Contributors*: 3idiot, A bit iffy, Abu badali, Abune, Acogbill, Activationensues, Agencius, Airwolf, Albanaco, AmadeoV, Antonrojo, ArielGold, Asparagirl, Atrian, Averyhse, BIL, BWH76, Ball of pain, BanyanTree, Barrymyles, Beano, Beeromatic, Bender235, Beyazid, BhaiSaab, Bobo192, Bovineone, Bricktop, Brucevdk, Burzum, CamperStrike, Can't sleep, clown will eat me, Cchris, Cczq, Chandra little, ChrisCork, Chuckhoffmann, Cnyborg, Conscious, Contributive, Crumley, Cyde, D.Kurdistani, DHR, DMCer, DMG413, DavidWBrooks, Derlay, Dhartung, Dismas, Downtownee, E71, Earth, Edward, Engology, Espace, Ferrisj, Feydey, Finn-Zoltan, Francis45, French user, GW Simulations, Gbabywiki, Gbleem, Glasser, Godardesque, Goldorack, Gurch, Gwernol, Hagerman, Havashavas, Hektor, Hooriaj, Hux, IRelayer, Ido50, Implificator, Inoculatedcities, Irooni hastam, Jake@prodea.com, Jennifer's Username, Jeresig, Jergen, Jmsilverstein, John, Johnisagoodman565, Joojoo ra, Jrockley, Jwillbur, KHerbst, Kashk, Kaveh, Khorshid, Khosrow II, Kitch, Kostisl, Kourosh ziabari, Kragen, Krich, Kurdo777, Leesonma, Li-sung, LiberalFascist, Lihaas, LilHelpa, MBK004, MachoCarioca, Man2man, ManiF, Marmoulak, Maryam1104, Merchbow, Mesgary, Messiahxi, Mfield, Mineo3, Mirmehrabi, Miss-simworld, Mlpearc, MoRsE, Mondo6, Mrlukeplease, Natrajdr, Neftchi, Negrulio, Nepaheshgar, Neum, Nevit, Newtonke, Nixer, Nthdegx, Oneworld25, Onsly, Ovi ace, Parishan, Patchouli, PatrickFisher, Pejman47, Peteharding, Philip Stevens, Plasticup, Pmanderson, Pnazemi, Postcard Cathy, Python eggs, Qrc2006, Quinn56, Qworty, RadioFan, RandomCritic, Randy2063, Rclavenna, Reinoutr, Rillian, Rjm656s, Rjwilmsi, Rmhermen, Rob Hinks, Roozbeh, Sabzaby, Saeed, Sagqs, Sam8, Sanbec, Sanchom, Sangak, Sangak2, Scolaire, Sdsds, Shauheen, Shervinafshar, ShimaKatase, Shreevatsa, Siba, Sidatio, SimonMenashy, SimonP, Sina, SiobhanHansa, Slmcom, Smoth 007, Steel, Straightpress, StuffOfInterest, Suckstobeyou, Sverdrup, TH-Foreigner, Taka2007, The Behnam, The Lake Effect, The wub, Thumperward, Ticktocktic, TimDM85, Tom harrison, Tpayne, Tuncrypt, Ulric1313, Vanka5, Vesal, Vice regent, Vocaro, Voice of All, Vonones, Wahabijaz, Wilde2, Woohookitty, ZK001, Zereshk, Zippanova, Zootsuits, 318 anonymous edits

Charles Simonyi *Source*: http://en.wikipedia.org/w/index.php?oldid=364525488 *Contributors*: AbuMusab, Adam78, Alan Peakall, Anders.Warga, Angelbo, Anton Khorev, Apankrat, Arado, Asdirk, Astromp, Avono, AxelBoldt, AxiomShell, AyaK, Azazell0, Bandpass, Bevo, Blainster, Bletch, Blowdart, Borgx, Can't sleep, clown will eat me, Casper Gutman, Chympy, Clydecoast, Coelacan, Colds7ream, Crocodile Punter, D6, Dagoli, Dan100, DarthVader, Darwinek, David Gerard, David Kernow, Dejin, Dhartung, Diotti, Dojarca, Domminico, Donreed, Dpbsmith, Droplet3, Dwightwebber, Dzahler, EagleStand, Eh Nonymous, Ehn, Espace, Fdnythebravest, Findaknow, GW Simulations, Gamaliel, Gazpacho, Gbrandt, GoingBatty, Goliza, Gundam Bass, Gymboy689, Hede2000, Hektor, Hillman, Hobartimus, Huby1, Ilikepie2221, Ixfd64, JaGa, Jef-Infojef, Jefmom, Jeltz, Jhf, Jiang, Jj137, Jnc, John Vandenberg, Johnisagoodman565, Johs, Joy, Juliancolton, Kenyon, Kernel Saunters, Kharker, Krm500, Kubigula, Livingrm, Lord.lucan, Lsstteam, Luckyluke, MBK004, Macintosh User, Marcika, Masterpiece2000, Mayumashu, Mckoss, Mervyn, Miles1228, Mmarci, Mrwojo, Murkygrom, Natalie West, Northgrove, Obafgkm, Paulus78, Peruvianllama, Philip Cross, Philip Stevens, Rainer Wasserfuhr, Rich Farmbrough, Rillian, Roastytoast, Ronark, Runtime, Ruszewski, S Carpenter, Sakhalinrf, Sam8, Savolya, Sdsds, SeattleArch, Seegoon, Sj, SocratesJedi, Soniatbolt, Squash Racket, Ssilvers, StN, Starbuzz3d, Thumperward, Timc, TotoBaggins, Twp, Vegaswikian, W-dahl, WAS 4.250, Whereizben, Woohookitty, YUL89YYZ, Zamphuor, Zeacho, 161 anonymous edits

Richard Garriott *Source*: http://en.wikipedia.org/w/index.php?oldid=364945147 *Contributors*: -Majestic-, 84user, AlmostReadytoFly, Alus, Amanaplanacanalpanama, Anetode, ArielGold, Arnoox, Ashmoo, Bsimmons666, CanadianLinuxUser, Centrx, Chaos5023, CheMechanical, Chilin, Chill Pill Bill, Conti, Coolhawks88, Crankedup, Cretanforever, Cujozz, CyberSkull, D6, DShamen, DamsonDragon, Dancter, Debresser, Denny, Dmosher, DoubleBlue, Elonka, Ettrig, Florentino floro, Frecklefoot, GCW50, Gamma, GlassCobra, Gshock91, HarryHenryGebel, Hektor, HyperCapitalist, JFG, James.Portnow, Jatkins, Jimfbleak, John Doe or Jane Doe, JohnArmagh, JohnnyMrNinja, Johntex, Jpatokal, JustAGal, KC6ROL, Kariteh, Kitch, Koveras, LeoNomis, LorenzoB, MBK004, MJGHavoc22, Malcolm Farmer, Maralia, Melaen, MercZ, Mintleaf, Mrwojo, MuzikJunky, Nar Matteru, NeuronExMachina, Nifboy, Nihonjoe, Nowheresville, Onsly, OttoMäkelä, Pd THOR, Peteharding, Platypus222, Primogen, RN, Rich Farmbrough, Richhaddon, Rillian, Rjwilmsi, Rmatney, Rmhermen, Ronhjones, Rtalaman, Ryan Roos, SaintDaveUK, Sdsds, Shp0ng1e, Skybum, Smartse, Snigbrook, Spazik007, Stabby Joe, Stepa, Steveprutz, Stoo, Sydius, TPIRFanSteve, Tdadamemd, TeonEdwards, TerokNor, ThaddeusB, Thapthim, The Thing That Should Not Be, Theserialcomma, TravelingCat, Trhickman, Twp, Uli, UnlimitedAccess, Varsindarkin, Whoosher, Wilson44691, Wwwwolf, Xiliquiern, Yorick8080, ZFGokuSSJ1, ZPM, Zippanova, 149 anonymous edits

Guy Laliberté *Source*: http://en.wikipedia.org/w/index.php?oldid=364429021 *Contributors*: 2005, 6mat1, Adamjwinter, Ageekgal, AlainV, All Hallow's Wraith, Alm93, Atif.t2, Awiseman, Balloonman, Billinghurst, Bluetooth954, Bupbin, Can't sleep, clown will eat me, Canada Jack, Carabinieri, Conical Johnson, Connormah, D6, Debresser, DegenFarang, DeltaGoodremFan, Discospinster, Doyleb23, Endgame1, EnzoRoyale, Gaelic Rules, Ganeshk, Gary King, Gary Kirk, Good Olfactory, Grandmasterka, Hektor, HordeFTL, Ibagli, Igoulet, Iluvalar, Ixfd64, Janmarques, Japanese Searobin, Jeffrey O. Gustafson, Jonymamilou, Kappa, Kingboyk, Lilac Soul, Lyverbe, MBK004, MTLskyline, Mayumashu, Melnikov, Munchaussen, Nelson Fernandez, Peacecount, Petercimon, Piccadilly, RODERICKMOLASAR, Radagast, Recognizance, Rillian, Rjwilmsi, RoyBoy, SGGH, Schoen, Senseitaco, Sirex98, Slakr, Sonia, Sysys, Tdadamemd, ThePlaz, Thethinredline, Timwhit, Tjoneslo, Tregoweth, Trusilver, Unitfloat, V8Cougar, Vegaswikian, Vox Humana 8', Wizardman, YUL89YYZ, Yansa, 186 anonymous edits

Private spaceflight *Source*: http://en.wikipedia.org/w/index.php?oldid=368680611 *Contributors*: ATD, Aces lead, Alba, AlexKM, Andrew Swallow, Andykass, Antandrus, Anxietycello, AzaToth, Balderdash707, Benbest, Bob A, Bobblehead, Bobblewik, Bovineone, Brockert, CMBJ, Chad Vander Veen, ChrisO, Claymorgan, Closedmouth, CommonsDelinker, Conscious, CryptoQuick, Darklilac, Dateapp, Dawnseeker2000, Eluchil404, Eric Wester, Fplay, Frankieroberto, Fredrik, GW Simulations, Garik, Grant76, Hektor, Hu12, Huntster, Hux, IanOsgood, Ingolfson, JHuwaldt, JacobSheehy, Jaganath, Jatkins, Jonahtrainer, Jondaily, Joseph Solis in Australia, Justice for All, Karada, Kavanagh, Krupenya, Kynereth, Lawman, Legotech, Mac, Man on the Mars, Mboverload, MioUzaki, Mlm42, Moomint, Morven, N2e, Nasa-verve, NeuronExMachina, Noclevername, Ondon, Potatoswatter, QuantumEleven, Richyricardo 2000, Rjwilmsi, Rmhermen, SDC, Schmloof, Sdsds, ShimaKatase, Simultaneous, Soosed, SpaceTycoon, Spitwater, StuffOfInterest, Sulgran, That Guy, From That Show!, Tristan705, Ultramarine, Wikianon, Wolfkeeper, Wronkiew, WurmWoode, 98 anonymous edits

SpaceShipOne *Source*: http://en.wikipedia.org/w/index.php?oldid=362936109 *Contributors*: Aces lead, Achmelvic, Adelphos, Adrian.benko, Ahoerstemeier, Alex Klotz, Allstarecho, Amontero, Andy120290, Anthon.Eff, Anthony5429, Ares2009, Arminius, Asparagus, BTTNext, Bbpen, Bevo, Bevo873, Blokhead, Bobblewik, Bollinger, Bovineone, Brockert, Bryan Derksen, Bugan, Burgundavia, CesarB, ChaosR, Chick Bowen, Chinakow, Chinju, Chmod007, Chris 73, Chrisblore, Chrisjj, Chrisvls, Cmprince, Colin Marquardt, Cr85r22, Crashtools2, Crei0, CryptoDerk, Cyclomedia, Dale Arnett, Daniel Hooper, Daren, Darklilac, Darksun, Darth Panda, Datkins2, Dawkeye, DerHexer, DirkvdM, Doradus, Dpm64, DragonflySixtyseven, EDG161, Ed g2s, EdH, Edward, Eisnel, ElrondHubble, Eluchil404, Emerson7, Ericg, F15x28, FellGleaming, FleaPlus, Fnlayson, GaeusOctavius, Gbleem, Gdavidp, Gene Nygaard, Geoffrey, Geoffrey.landis, Golbez, Greyengine5, Gsl, Gwernol, Henry W. Schmitt, Hux, IanOsgood, Iceberg3k, Ikluft, India, Inj, Inputpersona, Jiang, Jmcc150, JohnOwens, Johnteslade, Johntex, Jorfer, Joseaperez, JoshuaZ, Jrockley, Jwolfe, Kaare, Kazvorpal, Kbdank71, Ke4roh, Keith Edkins, Kim for sure, Kiore, Kitch, Kizor, Kowloonese, Kungfuadam, Laguna72, Leif, Liftarn, Lommer, MarcK, Mateo LeFou, Michael Snow, Miketwo, Millosh, Mstrzerg, Mulad, My Cat inn, N2e, N328KF, Narayansg, Neilbeach, NickCatal, Nickj, Nneonneo, Nonagonal Spider, Okwestern, Omicron91, Omphaloscope, Opkoad, Patrick, Pearle, PedroPVZ, Peter Ellis, Phil Boswell, Philip Stevens, Posix memalign, QueenofBattle, Quintin3265, R'son-W, RadicalBender, Raerth, RazorICE, Rchandra, RedBLACKandBURN, RedWolf, Renegadeaven, Reubenbarton, Rillian, Rlandmann, Rmhermen, Rogerd, Ropers, RuneSylvester, RxS, SDC, SOMER, Sade, Sagaciousuk, Sardanaphalus, Saschaheim, Scambob, SchuminWeb, Scott Burley, Seabhcan, Seth Ilys, Shuki, SimonKSK, Simone, Slysplace, Smack, Solitude, Spacegoat, Spm, Staz69uk, Stefan, Steinmb, Steve Leach, Stuart Morrow, Superborsuk, Sverdrup, Szajd, Talk, Tellyaddict, TexasDex, Texture, The Anome, The Singing Badger, The Wookieepedian, TheFeds, Theo10011, ThreeBlindMice, Tim Bell, Timc, Tobyw87, Tone, Trekphiler, Trevor MacInnis, Vanka5, Vegaswikian, Vesta, WPPilot, Wikibob, Wolfkeeper, Xmnemonic, Yosri, Yuriybrisk, Zenithpress, ZephyrAnycon, ZnU, کشرز, 313 anonymous edits

Mike Melvill *Source*: http://en.wikipedia.org/w/index.php?oldid=358441225 *Contributors*: Ahoerstemeier, Akradecki, Alexmcfire, Aliekens, Avatar, Bgraabek, Bill Thayer, Binksternet, Bkonrad, Cardsplayer4life, Cconrad, Cyan22, D6, Dhaluza, Doradus, DubaiTerminator, Edcolins, Elf-friend, Freakofnurture, Hektor, Hooperbloob, Jatkins, Jef-Infojef, Jmcc150, Johnstone, Joseph Solis in Australia, Jrockley, Latitude0116, Logawi, Luna Santin, MBK004, Mecil, N328KF, N5iln, Nehrams2020, NeuronExMachina, Nixer, Noisy, Owen, PPGMD, Palnatoke, Panser Born, Philip Stevens, Pigsonthewing, Raker, Rillian, Rjwilmsi, SGT141, Seth Ilys, ShimaKatase, WAAtkins, WPPilot, Wikibob, Wikigi, 35 anonymous edits

Space Adventures *Source*: http://en.wikipedia.org/w/index.php?oldid=367217280 *Contributors*: Aces lead, Advantecon, Ageekgal, AndersFeder, ArielGold, Benvogel, Bobak, Bovineone, BrownHairedGirl, Bryan Derksen, Creidieki, Daniel J. Leivick, DavidWBrooks, Davodd, EconomicsGuy, Ed629, Edward, Esn, Fearless editor, Feyrauth, Fragglet, Frecklefoot, Ganninia, Hektor, Igny, Jay794, Jrockley, Juliancolton, Lilac Soul, MarkRose, Medic463, Mr. Billion, Mushroom, N2e, Nairebis, Nasa-verve, Nat682, Ndssguy, Old Guard, Patrick, Plasticup, Pontuse, QuantumG, Quarl, Raerth, Rjwilmsi, Rmhermen, Scog, Shibidee, ShimaKatase, SiobhanHansa, Spacecomm, Takarada, Tango, Tdietz1, Tempshill, Tenebrae, TerraFrost, Theserialcomma, Toytown Mafia, Tyrol5, Vegaswikian, Verad, WacoJacko, Wernher, Wjfox2005, Woohookitty, Xandi, Xnatedawgx, Xxxxxxxxxx, Zyxwfgh127, 97 anonymous edits

Virgin Galactic *Source*: http://en.wikipedia.org/w/index.php?oldid=368676540 *Contributors*: Aces lead, Achmelvic, Adamnaylor, Aerugo, Akradecki, Alex537, AlexandrDmitri, Alexgt, Alexmcfire, Anville, Ardfern, ArielGold, Arnauldvm, BMello1618, BOARshevik, Barstudent, BassBone, Belgarath TS, Benbest, Benvogel, Bettia, Bibliophylax, Blazotron, Bodyn, Bovineone, CR85747, Chris Edgemon, Chuq, Cleared as filed, Colds7ream, Cornince, Csmiller, Cybercobra, Czerka Corp., D-Katana, Darin-0, Dawnseeker2000, DeadEyeArrow, Dmh, Doradus, Dusti, Dustin Pearson, Eliyahu S, Epbr123, Ergzay, Extraordinary, Florentino floro, Fosnez, GW Simulations, Garrett.Curley, Gilliam, GrantHenninger, Guyjohnston, Hack-Man, Hairy Dude, Hearth, Hektor, Hsan22, Hullubulloo, Huntster, Huwbutts, IDC, Imagi-King, Immunize, Inetpup, Ingolfson, Itsnoteasy07, JHuwaldt, Jakash, Jatkins, Jdavidb, Jmartin678, Jonathon Black, Jrockley, Just64helpin, KAMiKAZOW, Kbrian, Kim Meyrick, Kingdon, Kirinsilva, Korossyl, Krellis, KriZe, Latitude0116, Ligulem, Lilath, Lkt1126, Lost puppies, MCB, MER-C, MJBurrage, MKawasaki, Mareino, Medic463, Mephiston999, Merzbow, Metron4, Metstotop333, Mhking, MiG, Midnighttommo, MilkMiruku, Mindeye, Mischv, Mithridates, Mmeijeri, Moondyne, N2e, N328KF, NeuronExMachina, Nickshanks, Olyus, Pattapattasplodge, PaulHanson, Pfahlstrom, Philcha, Philip Stevens, PhilipO, Philwelch, Pisomojado, Plexust, Plugwash, PoleyDee, Poleydee,

Pontuse, Pouya, PowerCS, PureLegend, Pyrrhus16, Quarl, Qwghlm, RadicalBender, Raerth, Rchamberlain, Richard Branson, Richhaddon, Rillian, Rjwilmsi, RobertWalden, Rst20xx, Runningman85, Rwxrwxrwx, Ryohazuki1987, SCEhardt, SU Linguist, Savidan, Schmloof, ScienceApologist, Sdsds, Seabhcan, Seth Ilys, ShimaKatase, Silverxxx, Silvonen, SkeletorUK, Skizzik, Skybum, Skymist, Softssa, SquallLeonhart ITA, Stephenchou0722, Stevenem, Strangethingintheland, Suede, Sushi brobbo, Talskiddy, Tere naam, Themanwithoutapast, Thumperward, Tide rolls, Tim Long, Tony Sidaway, Trident13, U5K0, UniverseToday, VampWillow, Vitaleyes, Wackymacs, Wangi, Washi, Webkami, Wikianon, Wikibob, Wizardimps, Xaliqen, Xcom888, Xxxxxxxxxxx, Yellowdesk, Yonatan, Zomic13, 大西洋鮭, 335 anonymous edits

Bigelow Aerospace *Source*: http://en.wikipedia.org/w/index.php?oldid=368767046 *Contributors*: Aces lead, Alsocal, AndreniW, Andrew Swallow, Anthonares, Anykey, Apoc2400, Apyule, Ardfern, Aremisasling, Arichar6, Augustz, Benbest, BigEdit, Bovineone, Brockert, Bryan Derksen, Dahveed323, Dalf, Dawnseeker2000, Dmh, Dschmelzer, Farras Octara, Gene Nygaard, Hektor, Hiberniantears, HomersGhost, Honeplus, Huntster, IanOsgood, Inputpersona, Iridescent, James Doehring, Jrockley, Kaiba, Karl1263, LemRay, Livajo, M, Majorkev, Mmeijeri, Mosca, N2e, N328KF, Ndoro, Neg2led, NeuronExMachina, Numbo3, Oleg Alexandrov, Pah246, Pellrexx, Phaidros81, Plasticup, Raerth, RedBLACKandBURN, Rich Farmbrough, Ricnun, Rjwilmsi, Rmhermen, Robtj966, Rory096, RubenGarciaHernandez, Sanmichele, Sharkface217, ShimaKatase, Shimgray, SpaceCaptain, Spitwater, Squibman, StuffOfInterest, ThreeBlindMice, Thue, Trebuchetguy, Vegaswikian, VitriolUK, WildOne, Wtfk, Xaliqen, Zoggie50, 88 anonymous edits

Space Island Group *Source*: http://en.wikipedia.org/w/index.php?oldid=126810545 *Contributors*: Bhludzin, No1lakersfan, Skysmith

Space Island Project *Source*: http://en.wikipedia.org/w/index.php?oldid=233388412 *Contributors*: Bhludzin, BirgitteSB, Leolaursen, Martinlissmyr, Rjwilmsi, Tainter, 2 anonymous edits

Space Tourism Society *Source*: http://en.wikipedia.org/w/index.php?oldid=328916678 *Contributors*: ArielGold, Elonka, Jrockley, Nevit, OLEF641, Tom walker, 4 anonymous edits

Image Sources, Licenses and Contributors

Image:Atmosphere layers-en.svg *Source*: http://en.wikipedia.org/w/index.php?title=File:Atmosphere_layers-en.svg *License*: unknown *Contributors*: NOAA & w:User:MysidUser:Mysid

Image:Aurora-SpaceShuttle-EO.jpg *Source*: http://en.wikipedia.org/w/index.php?title=File:Aurora-SpaceShuttle-EO.jpg *License*: Public Domain *Contributors*: NASA (Crew of STS-39)(Original uploader was Seth Ilys at en.wikipedia)

Image:Gemeni4 EVA.jpg *Source*: http://en.wikipedia.org/w/index.php?title=File:Gemeni4_EVA.jpg *License*: Public Domain *Contributors*: Interiot, 1 anonymous edits

File:US Space Walk 1967 Issue-5c.jpg *Source*: http://en.wikipedia.org/w/index.php?title=File:US_Space_Walk_1967_Issue-5c.jpg *License*: Public Domain *Contributors*: US Post Office

Image:Gagarin space suite.jpg *Source*: http://en.wikipedia.org/w/index.php?title=File:Gagarin_space_suite.jpg *License*: Attribution *Contributors*: Original uploader was Irpen at en.wikipedia

File:Aldrin Apollo 11.jpg *Source*: http://en.wikipedia.org/w/index.php?title=File:Aldrin_Apollo_11.jpg *License*: Public Domain *Contributors*: NASA

File:Human spaceflight.svg *Source*: http://en.wikipedia.org/w/index.php?title=File:Human_spaceflight.svg *License*: GNU Free Documentation License *Contributors*: User:NuclearVacuum

File:Flag of the Soviet Union.svg *Source*: http://en.wikipedia.org/w/index.php?title=File:Flag_of_the_Soviet_Union.svg *License*: Public Domain *Contributors*: A1, Alex Smotrov, Alvis Jean, BagnoHax, Denniss, EugeneZelenko, F l a n k e r, Fred J, G.dallorto, Garynysmon, Herbythyme, Homo lupus, Jake Wartenberg, MaggotMaster, Ms2ger, Nightstallion, Pianist, R-41, Rainforest tropicana, Sebyugez, Solbris, Storkk, Str4nd, Tabasco, ThomasPusch, Toben, Zscout370, Сepп, 55 anonymous edits

File:Flag of the United States.svg *Source*: http://en.wikipedia.org/w/index.php?title=File:Flag_of_the_United_States.svg *License*: Public Domain *Contributors*: User:Dbenbenn, User:Indolences, User:Jacobolus, User:Technion, User:Zscout370

File:Flag of the People's Republic of China.svg *Source*: http://en.wikipedia.org/w/index.php?title=File:Flag_of_the_People's_Republic_of_China.svg *License*: Public Domain *Contributors*: User:Denelson83, User:SKopp, User:Shizhao, User:Zscout370

File:Flag of Europe.svg *Source*: http://en.wikipedia.org/w/index.php?title=File:Flag_of_Europe.svg *License*: Public Domain *Contributors*: User:-xfi-, User:Dbenbenn, User:Funakoshi, User:Jeltz, User:Nightstallion, User:Paddu, User:Verdy p, User:Zscout370

File:Flag of Iraq, 1991-2004.svg *Source*: http://en.wikipedia.org/w/index.php?title=File:Flag_of_Iraq,_1991-2004.svg *License*: Public Domain *Contributors*: Burts, Dbenbenn, Dinsdagskind, J. Patrick Fischer, Mattes, R-41

File:Flag of Japan.svg *Source*: http://en.wikipedia.org/w/index.php?title=File:Flag_of_Japan.svg *License*: Public Domain *Contributors*: Various

File:Flag of India.svg *Source*: http://en.wikipedia.org/w/index.php?title=File:Flag_of_India.svg *License*: Public Domain *Contributors*: User:SKopp

File:Flag of Iran.svg *Source*: http://en.wikipedia.org/w/index.php?title=File:Flag_of_Iran.svg *License*: unknown *Contributors*: Various

File:Flag of Romania.svg *Source*: http://en.wikipedia.org/w/index.php?title=File:Flag_of_Romania.svg *License*: Public Domain *Contributors*: AdiJapan

Image:Stanford torus under construction.jpg *Source*: http://en.wikipedia.org/w/index.php?title=File:Stanford_torus_under_construction.jpg *License*: Public Domain *Contributors*: w:Don Davis (artist)Don Davis

Image:An Experiment on a Bird in an Air Pump by Joseph Wright of Derby, 1768.jpg *Source*: http://en.wikipedia.org/w/index.php?title=File:An_Experiment_on_a_Bird_in_an_Air_Pump_by_Joseph_Wright_of_Derby,_1768.jpg *License*: Public Domain *Contributors*: Anne97432, AnonMoos, Duesentrieb, GeeJo, Gryffindor, Mattes, Micheletb, Pieter Kuiper, Pmsyyz, Rocket000, WikipediaMaster

Image:Moonmir sts91 big.jpg *Source*: http://en.wikipedia.org/w/index.php?title=File:Moonmir_sts91_big.jpg *License*: Public Domain *Contributors*: NASA

Image:Space fluid shift.gif *Source*: http://en.wikipedia.org/w/index.php?title=File:Space_fluid_shift.gif *License*: Public Domain *Contributors*: Athaenara, Gengiskanhg, Marcok, 3 anonymous edits

File:Space adaptation syndrome acclimation.jpg *Source*: http://en.wikipedia.org/w/index.php?title=File:Space_adaptation_syndrome_acclimation.jpg *License*: Public Domain *Contributors*: NASA's Marshall Space Flight Center and Science@NASA (U.S. Federal Government)

File:Mark Shuttleworth NASA.jpg *Source*: http://en.wikipedia.org/w/index.php?title=File:Mark_Shuttleworth_NASA.jpg *License*: Public Domain *Contributors*: Dbenbenn, Nikai, Startaq

File:Ansari X-Prize Check.jpg *Source*: http://en.wikipedia.org/w/index.php?title=File:Ansari_X-Prize_Check.jpg *License*: GNU Free Documentation License *Contributors*: Belb, Foroa, Ingolfson, Kbh3rd, Maksim, Wikipeder, 1 anonymous edits

File:Spaceship One in flight 1.jpg *Source*: http://en.wikipedia.org/w/index.php?title=File:Spaceship_One_in_flight_1.jpg *License*: Creative Commons Attribution-Sharealike 1.0 *Contributors*: Chris 73, Dbenbenn, Denniss, Medium69

File:Flag of South Africa.svg *Source*: http://en.wikipedia.org/w/index.php?title=File:Flag_of_South_Africa.svg *License*: Public Domain *Contributors*: User:SKopp

File:Flag of Hungary.svg *Source*: http://en.wikipedia.org/w/index.php?title=File:Flag_of_Hungary.svg *License*: Public Domain *Contributors*: User:SKopp

File:Flag of Malaysia.svg *Source*: http://en.wikipedia.org/w/index.php?title=File:Flag_of_Malaysia.svg *License*: Public Domain *Contributors*: User:SKopp

File:Flag of South Korea.svg *Source*: http://en.wikipedia.org/w/index.php?title=File:Flag_of_South_Korea.svg *License*: Public Domain *Contributors*: Various

File:Flag of Canada.svg *Source*: http://en.wikipedia.org/w/index.php?title=File:Flag_of_Canada.svg *License*: Public Domain *Contributors*: User:E Pluribus Anthony, User:Mzajac

File:Flag of Russia.svg *Source*: http://en.wikipedia.org/w/index.php?title=File:Flag_of_Russia.svg *License*: Public Domain *Contributors*: AndriusG, Artem Karimov, Davepape, Dmitry Strotsev, Drieskamp, Enbéká, Fred J, Gleb Borisov, Herbythyme, Homo lupus, Kiensvay, Klemen Kocjancic, Kwj2772, Mattes, Maximaximax, Miyokan, Nightstallion, Ondřej Žváček, Pianist, Pumbaa80, Putnik, R-41, Radziun, Rainman, Reisio, Rfc1394, Rkt2312, Rocket000, Sasa Stefanovic, SeNeKa, Srtxg, Stianbh, Wikiborg, Winterheart, Zscout370, Zyido, ОйЛ, 34 anonymous edits

File:Flag of the United Kingdom.svg *Source*: http://en.wikipedia.org/w/index.php?title=File:Flag_of_the_United_Kingdom.svg *License*: Public Domain *Contributors*: User:Zscout370

Image:Soyuz TMA-7 spacecraft2edit1.jpg *Source*: http://en.wikipedia.org/w/index.php?title=File:Soyuz_TMA-7_spacecraft2edit1.jpg *License*: Public Domain *Contributors*: Original uploader was Thegreenj at en.wikipedia

Image:Soyuz diagrama.gif *Source*: http://en.wikipedia.org/w/index.php?title=File:Soyuz_diagrama.gif *License*: Public Domain *Contributors*: Bryan Derksen, Campani, Pline, Will Pittenger, 1 anonymous edits

Image:Soyuz-TMA parts.jpg *Source*: http://en.wikipedia.org/w/index.php?title=File:Soyuz-TMA_parts.jpg *License*: Public Domain *Contributors*: NASA

Image:Soyuz-TMA orbital module.jpg *Source*: http://en.wikipedia.org/w/index.php?title=File:Soyuz-TMA_orbital_module.jpg *License*: Public Domain *Contributors*: NASA

Image:Soyuz-TMA descent module.jpg *Source*: http://en.wikipedia.org/w/index.php?title=File:Soyuz-TMA_descent_module.jpg *License*: Public Domain *Contributors*: NASA

Image:Soyuz-TMA propulsion module.jpg *Source*: http://en.wikipedia.org/w/index.php?title=File:Soyuz-TMA_propulsion_module.jpg *License*: Public Domain *Contributors*: NASA

Image:SOYUZ.png *Source*: http://en.wikipedia.org/w/index.php?title=File:SOYUZ.png *License*: Public Domain *Contributors*: User:Ricnun

Image:Soyuz 7K-T 3-seats drawing.png *Source*: http://en.wikipedia.org/w/index.php?title=File:Soyuz_7K-T_3-seats_drawing.png *License*: Public Domain *Contributors*: NASA

Image:Soyuz-T drawing.svg *Source*: http://en.wikipedia.org/w/index.php?title=File:Soyuz-T_drawing.svg *License*: Public Domain *Contributors*: NASA

Image:Soyuz-TM drawing.png *Source*: http://en.wikipedia.org/w/index.php?title=File:Soyuz-TM_drawing.png *License*: Public Domain *Contributors*: NASA

Image:Soyuz TMA-6 spacecraft.jpg *Source*: http://en.wikipedia.org/w/index.php?title=File:Soyuz_TMA-6_spacecraft.jpg *License*: Public Domain *Contributors*: NASA

Image:Soyuz National Space Centre.jpg *Source*: http://en.wikipedia.org/w/index.php?title=File:Soyuz_National_Space_Centre.jpg *License*: GNU Free Documentation License *Contributors*: Hellisp

Image:ASTP Soyuz Spacecraft.jpg *Source*: http://en.wikipedia.org/w/index.php?title=File:ASTP_Soyuz_Spacecraft.jpg *License*: unknown *Contributors*: Bryan Derksen, Edward, Hellisp, High Contrast, Red devil 666

Image:Soyuz 19 (Apollo Soyuz Test Project) spacecraft.jpg *Source*: http://en.wikipedia.org/w/index.php?title=File:Soyuz_19_(Apollo_Soyuz_Test_Project)_spacecraft.jpg *License*: Public Domain *Contributors*: NASA

Image:Soyuz acoplada MIR.jpg *Source*: http://en.wikipedia.org/w/index.php?title=File:Soyuz_acoplada_MIR.jpg *License*: Public Domain *Contributors*: NASA

Image:Soyuz spacecraft.jpg *Source*: http://en.wikipedia.org/w/index.php?title=File:Soyuz_spacecraft.jpg *License*: Creative Commons Attribution-Sharealike 2.5 *Contributors*: Poppy

Image:Soyuz rocket engines.jpg *Source*: http://en.wikipedia.org/w/index.php?title=File:Soyuz_rocket_engines.jpg *License*: Public Domain *Contributors*: NASA/Bill Ingalls

Image:Soyuz tm-31 transported to launch pad.jpg *Source*: http://en.wikipedia.org/w/index.php?title=File:Soyuz_tm-31_transported_to_launch_pad.jpg *License*: Public Domain *Contributors*: NASA

Image:Soyuz TMA-2 launch.jpg *Source*: http://en.wikipedia.org/w/index.php?title=File:Soyuz_TMA-2_launch.jpg *License*: Public Domain *Contributors*: NASA/Scott Andrews

Image:Soyuz landing.jpg *Source*: http://en.wikipedia.org/w/index.php?title=File:Soyuz_landing.jpg *License*: Public Domain *Contributors*: Original uploader was Error 404 at en.wikipedia

Image:Space Shuttle vs Soyuz TM - to scale drawing.png *Source*: http://en.wikipedia.org/w/index.php?title=File:Space_Shuttle_vs_Soyuz_TM_-_to_scale_drawing.png *License*: Public Domain *Contributors*: NASA

Image:Soyus craft on Display at the Kennedy Space Center.jpg *Source*: http://en.wikipedia.org/w/index.php?title=File:Soyus_craft_on_Display_at_the_Kennedy_Space_Center.jpg *License*: Public Domain *Contributors*: Original uploader was Halibut Thyme at en.wikipedia

Image:SoyuzSovietStamp.jpg *Source*: http://en.wikipedia.org/w/index.php?title=File:SoyuzSovietStamp.jpg *License*: Public Domain *Contributors*: User:HordeFTL

Image:Jsc2006e33951.jpg *Source*: http://en.wikipedia.org/w/index.php?title=File:Jsc2006e33951.jpg *License*: Public Domain *Contributors*: NASA

Image:S132e012208.jpg *Source*: http://en.wikipedia.org/w/index.php?title=File:S132e012208.jpg *License*: Public Domain *Contributors*: NASA

Image:ISS Logo.svg *Source*: http://en.wikipedia.org/w/index.php?title=File:ISS_Logo.svg *License*: Public Domain *Contributors*: User:Thommy9

Image:ISS insignia.svg *Source*: http://en.wikipedia.org/w/index.php?title=File:ISS_insignia.svg *License*: Public Domain *Contributors*: Original: NASA This vector version by w:User:MysidMysid

Image:ISS configuration 2010-05 en.svg *Source*: http://en.wikipedia.org/w/index.php?title=File:ISS_configuration_2010-05_en.svg *License*: Public Domain *Contributors*: NASA and James Humphreys (Colds7ream)

File:183478main MSG1.jpg *Source*: http://en.wikipedia.org/w/index.php?title=File:183478main_MSG1.jpg *License*: Public Domain *Contributors*: NASA/Crew of Expedition 8

File:382009main MELFI12.jpg *Source*: http://en.wikipedia.org/w/index.php?title=File:382009main_MELFI12.jpg *License*: Public Domain *Contributors*: NASA

File:Atlantis Docked to Mir.jpg *Source*: http://en.wikipedia.org/w/index.php?title=File:Atlantis_Docked_to_Mir.jpg *License*: unknown *Contributors*: User:Jhs

File:STS-124 Garan EVA2.jpg *Source*: http://en.wikipedia.org/w/index.php?title=File:STS-124_Garan_EVA2.jpg *License*: Public Domain *Contributors*: NASA Original uploader was Utahredrock at en.wikipedia

File:Zarya from STS-88.jpg *Source*: http://en.wikipedia.org/w/index.php?title=File:Zarya_from_STS-88.jpg *License*: Public Domain *Contributors*: Bricktop, Nuno Tavares, Pline, TheDJ, 1 anonymous edits

File:ISS Unity module.jpg *Source*: http://en.wikipedia.org/w/index.php?title=File:ISS_Unity_module.jpg *License*: Public Domain *Contributors*: NASA

File:ISS Zvezda module-small.jpg *Source*: http://en.wikipedia.org/w/index.php?title=File:ISS_Zvezda_module-small.jpg *License*: Public Domain *Contributors*: NASA

File:ISS Destiny Lab.jpg *Source*: http://en.wikipedia.org/w/index.php?title=File:ISS_Destiny_Lab.jpg *License*: Public Domain *Contributors*: NASA

File:ISS Quest airlock.jpg *Source*: http://en.wikipedia.org/w/index.php?title=File:ISS_Quest_airlock.jpg *License*: Public Domain *Contributors*: NASA

File:Pirs docking module taken by STS-108.jpg *Source*: http://en.wikipedia.org/w/index.php?title=File:Pirs_docking_module_taken_by_STS-108.jpg *License*: Public Domain *Contributors*: NASA

File:Harmony Relocation.jpg *Source*: http://en.wikipedia.org/w/index.php?title=File:Harmony_Relocation.jpg *License*: Public Domain *Contributors*: NASA TV (Expedition 16 crew)

File:S122e007873.jpg *Source*: http://en.wikipedia.org/w/index.php?title=File:S122e007873.jpg *License*: Public Domain *Contributors*: NASA

File:Kibo ELM-PS on ISS.jpg *Source*: http://en.wikipedia.org/w/index.php?title=File:Kibo_ELM-PS_on_ISS.jpg *License*: Public Domain *Contributors*: NASA

File:STS-124 Kibo.jpg *Source*: http://en.wikipedia.org/w/index.php?title=File:STS-124_Kibo.jpg *License*: Public Domain *Contributors*: NASA

File:Poisk.Jpeg *Source*: http://en.wikipedia.org/w/index.php?title=File:Poisk.Jpeg *License*: Public Domain *Contributors*: NASA

File:Tranquility-node3.JPG *Source*: http://en.wikipedia.org/w/index.php?title=File:Tranquility-node3.JPG *License*: Public Domain *Contributors*: NASA

File:STS-130 Nicholas Patrick looks through Cupola.jpg *Source*: http://en.wikipedia.org/w/index.php?title=File:STS-130_Nicholas_Patrick_looks_through_Cupola.jpg *License*: Public Domain *Contributors*: NASA

File:Iss023e047527.jpg *Source*: http://en.wikipedia.org/w/index.php?title=File:Iss023e047527.jpg *License*: Public Domain *Contributors*: NASA/Federal Governemnt

File:STS-114 Raffaello module.jpg *Source*: http://en.wikipedia.org/w/index.php?title=File:STS-114_Raffaello_module.jpg *License*: Public Domain *Contributors*: NASA

File:MLM - ISS module.jpg *Source*: http://en.wikipedia.org/w/index.php?title=File:MLM_-_ISS_module.jpg *License*: unknown *Contributors*: NASA

File:ISS Crew Return Vehicle.jpg *Source*: http://en.wikipedia.org/w/index.php?title=File:ISS_Crew_Return_Vehicle.jpg *License*: Public Domain *Contributors*: Carla Thomas

File:STS-114 Steve Robinson on Canadarm2.jpg *Source*: http://en.wikipedia.org/w/index.php?title=File:STS-114_Steve_Robinson_on_Canadarm2.jpg *License*: Public Domain *Contributors*: NASA

File:ISS on 20 August 2001.jpg *Source*: http://en.wikipedia.org/w/index.php?title=File:ISS_on_20_August_2001.jpg *License*: Public Domain *Contributors*: NASA

File:Internationale Raumstation Bahnhöhe (dumb version).png *Source*: http://en.wikipedia.org/w/index.php?title=File:Internationale_Raumstation_Bahnhöhe_(dumb_version).png *License*: Public Domain *Contributors*: rmw. Original uploader was Rmw at de.wikipedia

File:ISS Communication Systems.png *Source*: http://en.wikipedia.org/w/index.php?title=File:ISS_Communication_Systems.png *License*: Public Domain *Contributors*: Gary Kitmacher/NASA

File:SpaceStationCycle.svg *Source*: http://en.wikipedia.org/w/index.php?title=File:SpaceStationCycle.svg *License*: Public Domain *Contributors*: User:AntiNeo

Image:ISS 26.07.07.jpg *Source*: http://en.wikipedia.org/w/index.php?title=File:ISS_26.07.07.jpg *License*: Public Domain *Contributors*: Andreas Möller aka Lug2k

File:ISS Main Contributors.svg *Source*: http://en.wikipedia.org/w/index.php?title=File:ISS_Main_Contributors.svg *License*: Public Domain *Contributors*: User:Emilfaro

File:ISS Hardware Allocation.png *Source*: http://en.wikipedia.org/w/index.php?title=File:ISS_Hardware_Allocation.png *License*: Creative Commons Attribution-Sharealike 3.0 *Contributors*: User:Colds7ream

File:Temporary Sleep Station.jpg *Source*: http://en.wikipedia.org/w/index.php?title=File:Temporary_Sleep_Station.jpg *License*: Public Domain *Contributors*: NASA

File:Meal STS127.jpg *Source*: http://en.wikipedia.org/w/index.php?title=File:Meal_STS127.jpg *License*: Public Domain *Contributors*: NASA

File:TVIS treadmill.jpg *Source*: http://en.wikipedia.org/w/index.php?title=File:TVIS_treadmill.jpg *License*: Public Domain *Contributors*: NASA/Crew of Expedition 14

File:STS-118 approaching ISS.jpg *Source*: http://en.wikipedia.org/w/index.php?title=File:STS-118_approaching_ISS.jpg *License*: Public Domain *Contributors*: NASA

File:ISS Centers.svg *Source*: http://en.wikipedia.org/w/index.php?title=File:ISS_Centers.svg *License*: Public Domain *Contributors*: w:en:User:TastyCakesTastyCakes

File:Magnify-clip.png *Source*: http://en.wikipedia.org/w/index.php?title=File:Magnify-clip.png *License*: GNU Free Documentation License *Contributors*: User:Erasoft24

File:STS-118 debris entry.jpg *Source*: http://en.wikipedia.org/w/index.php?title=File:STS-118_debris_entry.jpg *License*: Public Domain *Contributors*: NASA

Image:Dennis Tito.jpg *Source*: http://en.wikipedia.org/w/index.php?title=File:Dennis_Tito.jpg *License*: Public Domain *Contributors*: NASA

Image:Soyuz-tm32.jpg *Source*: http://en.wikipedia.org/w/index.php?title=File:Soyuz-tm32.jpg *License*: Public Domain *Contributors*: Unknown

Image:Mark Shuttleworth by Martin Schmitt.jpg *Source*: http://en.wikipedia.org/w/index.php?title=File:Mark_Shuttleworth_by_Martin_Schmitt.jpg *License*: Creative Commons Attribution 2.0 *Contributors*: Martin Schmitt (cropped by Mary Gardiner)

Image:Soyuz-tm34.gif *Source*: http://en.wikipedia.org/w/index.php?title=File:Soyuz-tm34.gif *License*: Public Domain *Contributors*: Enen, West

File:Mark-Shuttleworth-Ubuntu-fr-Karmic.jpg *Source*: http://en.wikipedia.org/w/index.php?title=File:Mark-Shuttleworth-Ubuntu-fr-Karmic.jpg *License*: Creative Commons Attribution-Sharealike 3.0 *Contributors*: User:Nitot

Image:Mark Shuttleworth NASA.jpg *Source*: http://en.wikipedia.org/w/index.php?title=File:Mark_Shuttleworth_NASA.jpg *License*: Public Domain *Contributors*: Dbenbenn, Nikai, Startaq

Image:Gregory Olsen.jpg *Source*: http://en.wikipedia.org/w/index.php?title=File:Gregory_Olsen.jpg *License*: Public Domain *Contributors*: Astromp, Bricktop, Schekinov Alexey Victorovich, Startaq, 1 anonymous edits

Image:Soyuz TMA-7 Patch.jpg *Source*: http://en.wikipedia.org/w/index.php?title=File:Soyuz_TMA-7_Patch.jpg *License*: Public Domain *Contributors*: Avron, Lotse

Image:Anousheh Ansari.jpg *Source*: http://en.wikipedia.org/w/index.php?title=File:Anousheh_Ansari.jpg *License*: Public Domain *Contributors*: NASA

Image:Soyuz TMA-9 Logo.jpg *Source*: http://en.wikipedia.org/w/index.php?title=File:Soyuz_TMA-9_Logo.jpg *License*: Public Domain *Contributors*: Avron, Lokal Profil, MachoCarioca, Mike.lifeguard, Panther, TheDJ

Image:Anousheh Ansari in the ISS.jpg *Source*: http://en.wikipedia.org/w/index.php?title=File:Anousheh_Ansari_in_the_ISS.jpg *License*: Public Domain *Contributors*: NASA

Image:Anousheh Ansari space patch.jpg *Source*: http://en.wikipedia.org/w/index.php?title=File:Anousheh_Ansari_space_patch.jpg *License*: unknown *Contributors*: X-Prize

Image:Soyuz TMA-9 crew w ansari.jpg *Source*: http://en.wikipedia.org/w/index.php?title=File:Soyuz_TMA-9_crew_w_ansari.jpg *License*: Public Domain *Contributors*: Victor Zelentsov/NASA

File:Charles simonyi.jpg *Source*: http://en.wikipedia.org/w/index.php?title=File:Charles_simonyi.jpg *License*: Public Domain *Contributors*: NASA

File:Richard garriott july 2008.jpg *Source*: http://en.wikipedia.org/w/index.php?title=File:Richard_garriott_july_2008.jpg *License*: Public Domain *Contributors*: NASA

Image:WinEarth screen Peru coast.png *Source*: http://en.wikipedia.org/w/index.php?title=File:WinEarth_screen_Peru_coast.png *License*: Public Domain *Contributors*: Windows on Earth, Dan Barstow, Teon Edwards

File:Guy_Laliberte_WPT.jpg *Source*: http://en.wikipedia.org/w/index.php?title=File:Guy_Laliberte_WPT.jpg *License*: Creative Commons Attribution-Sharealike 2.0 *Contributors*: Photos by flipchip / LasVegasVegas.com

Image:Laliberté Soyuz TMA16.jpg *Source*: http://en.wikipedia.org/w/index.php?title=File:Laliberté_Soyuz_TMA16.jpg *License*: Public Domain *Contributors*: NASA

File:Soyuz-TMA-16-Mission-Patch.png *Source*: http://en.wikipedia.org/w/index.php?title=File:Soyuz-TMA-16-Mission-Patch.png *License*: Public Domain *Contributors*: Anastasiya Mestyashova

Image:Satellites For Sale - GPN-2000-001036.jpg *Source*: http://en.wikipedia.org/w/index.php?title=File:Satellites_For_Sale_-_GPN-2000-001036.jpg *License*: Public Domain *Contributors*: NASA

Image:Proton Zvezda crop.jpg *Source*: http://en.wikipedia.org/w/index.php?title=File:Proton_Zvezda_crop.jpg *License*: Public Domain *Contributors*: NASA

File:Dragon ISS.jpg *Source*: http://en.wikipedia.org/w/index.php?title=File:Dragon_ISS.jpg *License*: Public Domain *Contributors*: NASA

Image:Spaceship One in flight 1.jpg *Source*: http://en.wikipedia.org/w/index.php?title=File:Spaceship_One_in_flight_1.jpg *License*: Creative Commons Attribution-Sharealike 1.0 *Contributors*: Chris 73, Dbenbenn, Denniss, Medium69

Image:X-33 Venture Star.jpg *Source*: http://en.wikipedia.org/w/index.php?title=File:X-33_Venture_Star.jpg *License*: Public Domain *Contributors*: NASA

Image:asteroidmining.jpg *Source*: http://en.wikipedia.org/w/index.php?title=File:Asteroidmining.jpg *License*: unknown *Contributors*: Kavanagh, Liftarn

Image: IMG 1237.jpg *Source*: http://en.wikipedia.org/w/index.php?title=File:IMG_1237.jpg *License*: GNU Free Documentation License *Contributors*: Dbenbenn, Ingolfson, Quadell, Renegadeaven, WPPilot

Image:SS1Group01.jpg *Source*: http://en.wikipedia.org/w/index.php?title=File:SS1Group01.jpg *License*: Creative Commons Attribution-Sharealike 3.0 *Contributors*: User:WPPilot

Image:Kluft-photo-SS1-landing-June-2004-Img 1406c.jpg *Source*: http://en.wikipedia.org/w/index.php?title=File:Kluft-photo-SS1-landing-June-2004-Img_1406c.jpg *License*: GNU Free Documentation License *Contributors*: User:Ikluft

Image:Spaceship One Other Angle.JPG *Source*: http://en.wikipedia.org/w/index.php?title=File:Spaceship_One_Other_Angle.JPG *License*: Public Domain *Contributors*: User:Jorfer

Image:SpaceShipOneReplica.JPG *Source*: http://en.wikipedia.org/w/index.php?title=File:SpaceShipOneReplica.JPG *License*: Free Art License *Contributors*: User:Elrondhubble

Image:Spaceship One cockpit in flight.jpg *Source*: http://en.wikipedia.org/w/index.php?title=File:Spaceship_One_cockpit_in_flight.jpg *License*: unknown *Contributors*: Akradecki, Chris 73, Ingolfson, KTo288, Stahlkocher, 1 anonymous edits

Image:Spaceship One and White Knight rear view.jpg *Source*: http://en.wikipedia.org/w/index.php?title=File:Spaceship_One_and_White_Knight_rear_view.jpg *License*: Creative Commons Attribution-Sharealike 1.0 *Contributors*: Chris 73, Denniss, Ingolfson

Image:SpaceShipOne schematic.png *Source*: http://en.wikipedia.org/w/index.php?title=File:SpaceShipOne_schematic.png *License*: Public Domain *Contributors*: User:Jrockley

Image:Ss1 smithsonian.jpg *Source*: http://en.wikipedia.org/w/index.php?title=File:Ss1_smithsonian.jpg *License*: GNU Free Documentation License *Contributors*: User ZnU on en.wikipedia

Image:Spaceship One.JPG *Source*: http://en.wikipedia.org/w/index.php?title=File:Spaceship_One.JPG *License*: Public Domain *Contributors*: User:Jorfer

Image:patti-presenting-wings-web.jpg *Source*: http://en.wikipedia.org/w/index.php?title=File:Patti-presenting-wings-web.jpg *License*: Public Domain *Contributors*: Original uploader was Nixer at en.wikipedia

File:US - FAA Astronaut Wings.png *Source*: http://en.wikipedia.org/w/index.php?title=File:US_-_FAA_Astronaut_Wings.png *License*: Public Domain *Contributors*: United States Federal Aviation Administration

Image:MikeandScottC01.jpg *Source*: http://en.wikipedia.org/w/index.php?title=File:MikeandScottC01.jpg *License*: Creative Commons Attribution-Sharealike 3.0 *Contributors*: User:WPPilot

File:Space adventures logo-clear.jpg *Source*: http://en.wikipedia.org/w/index.php?title=File:Space_adventures_logo-clear.jpg *License*: unknown *Contributors*: Tyrol5

Image:Virgin Galactic.png *Source*: http://en.wikipedia.org/w/index.php?title=File:Virgin_Galactic.png *License*: unknown *Contributors*: SkeletorUK

Image:Vg-wk2-080728 (148) cr8.jpg *Source*: http://en.wikipedia.org/w/index.php?title=File:Vg-wk2-080728_(148)_cr8.jpg *License*: Creative Commons Attribution-Sharealike 3.0 *Contributors*: User:Akradecki

Image:BigelowLogo.png *Source*: http://en.wikipedia.org/w/index.php?title=File:BigelowLogo.png *License*: unknown *Contributors*: Bigelow Aerospace

Image:NautilusModule2.jpg *Source*: http://en.wikipedia.org/w/index.php?title=File:NautilusModule2.jpg *License*: unknown *Contributors*: Bigelow Aerospace

Image:GenesisI.jpeg *Source*: http://en.wikipedia.org/w/index.php?title=File:GenesisI.jpeg *License*: unknown *Contributors*: Bigelow Aerospace

NU Free Documentation License Version 1.2, ovember 2002 Copyright (C) 2000,2001,2002 ee Software Foundation, Inc. 59 Temple lace, Suite 330, Boston, MA 02111 -1307 USA veryone is permitted to copy and distribute erbatim copies of this license documen t, but hanging it is not allowed.

PREAMBLE

he purpose of this License is to make a manual, textbook, or her functional and useful document "free" in the sense of edom: to assure everyone the effective freedom to copy and distribute it, with or without modifying it, either commercially or ncommercially. Secondarily, this License preserves for the thor and publisher a way to get credit for their work, while not ing considered responsible for modifications made by others. is License is a kind of "copyleft", which means that derivative rks of the document must themselves be free in the same nse. It complements the GNU General Public License, which is copyleft license designed for free software. We have designed s License in order to use it for manuals for free software, cause free software needs free documentation: a free program ould come with manuals providing the same freedoms that the ftware does. But this License is not limited to software manuals; an be used for any textual work, regardless of subject matter whether it is published as a printed book. We recommend this ense principally for works whose purpose is instruction or ference.

APPLICABILITY AND DEFINITIONS

his License applies to any manual or ot her work, in any edium, that contains a notice placed by the copyright holder ying it can be distributed under the terms of this License. Such notice grants a world -wide, royalty -free license, unlimited in ration, to use that work under the conditio ns stated herein. The ocument", below, refers to any such manual or work. Any ember of the public is a licensee, and is addressed as "you". u accept the license if you copy, modify or distribute the work a way requiring permission under copyright l aw. A "Modified rsion" of the Document means any work containing the cument or a portion of it, either copied verbatim, or with difications and/or translated into another language. A econdary Section" is a named appendix or a front -matter ction o f the Document that deals exclusively with the ationship of the publishers or authors of the Document to the cument's overall subject (or to related matters) and contains thing that could fall directly within that overall subject. (Thus, if Docu ment is in part a textbook of mathematics, a Secondary ction may not explain any mathematics.) The relationship could a matter of historical connection with the subject or with ated matters, or of legal, commercial, philosophical, ethical or litical position regarding them. The "Invariant Sections" are tain Secondary Sections whose titles are designated, as being ose of Invariant Sections, in the notice that says that the cument is released under this License. If a section does not fit above definition of Secondary then it is not allowed to be signated as Invariant. The Document may contain zero variant Sections. If the Document does not identify any Invariant ctions then there are none. The "Cover Texts" are certain short ssages of text that are listed, as Front -Cover Texts or Back - ver Texts, in the notice that says that the Document is eased under this License. A Front -Cover Text may be at most vords, and a Back -Cover Text may be at most 25 words. A ansparent" copy of the Document means a machine -readable py, represented in a format whose specification is available to general public, that is suitable for revising the document aightforwardly with generic text editors or (for images mposed of pixels) generic p aint programs or (for drawings) ne widely available drawing editor, and that is suitable for input text formatters or for automatic translation to a variety of mats suitable for input to text formatters. A copy made in an herwise Transparent file format whose markup, or absence of rkup, has been arranged to thwart or discourage subsequent dification by readers is not Transparent. An image format is t Transparent if used for any substantial amount of text. A copy at is not "Transparent" is called "Opaque". Examples of suitable mats for Transparent copies include plain ASCII without rkup, Texinfo input format, LaTeX input format, SGML or XML ng a publicly available DTD, and standard -conforming simple ML, PostScript or PDF designed f or human modification. amples of transparent image formats include PNG, XCF and G. Opaque formats include proprietary formats that can be ad and edited only by proprietary word processors, SGML or IL for which the DTD and/or processing tools are not generally ailable, and the machine -generated HTML, PostScript or PDF duced by some word processors for output purposes only. The tle Page" means, for a printed book, the title page itself, plus ch following pages as are needed to hold, legibly, the material License requires to appear in the title page. For works in mats which do not have any title page as such, "Title Page" ans the text near the most prominent appearance of the work's , preceding the beginning of the body of the te xt. A section ntitled XYZ" means a named subunit of the Document whose either is precisely XYZ or contains XYZ in parentheses owing text that translates XYZ in another language. (Here XYZ nds for a specific section name mentioned below, such as cknowledgements", "Dedications", "Endorsements", or story".) To "Preserve the Title" of such a section when you dify the Document means that it remains a section "Entitled Z" according to this definition. The Document may include rranty Disc laimers next to the notice which states that this ense applies to the Document. These Warranty Disclaimers considered to be included by reference in this License, but y as regards disclaiming warranties: any other implication that se Warranty Disclaimers may have is void and has no effect the meaning of this License.

VERBATIM COPYING

u may copy and distribute the Document in any medium, her commercially or noncommercially, provided that this ense, the copyright notices, and the license notice saying this License applies to the Document are reproduced in all copies, and that you add no other conditions whatsoever to those of this License. You may not use technical measures to obstruct or control the reading or further copying of t he copies you make or distribute. However, you may accept compensation in exchange for copies. If you distribute a large enough number of copies you must also follow the conditions in section 3. You may also lend copies, under the same conditions stated ab ove, and you may publicly display copies.

3. COPYING IN QUANTITY

If you publish printed copies (or copies in media that commonly have printed covers) of the Document, numbering more than 100, and the Document's license notice requires Cover Texts, you must enclose the copies in covers that carry, clearly and legibly, all these Cover Texts: Front -Cover Texts on the front cover, and Back-Cover Texts on the back cover. Both covers must also clearly and legibly identify you as the publisher of these copies. The front cover must present the full title with all words of the title equally prominent and visible. You may add other material on the covers in addition. Copying with changes limited to the covers, as long as they preserve the title of the Document and s atisfy these conditions, can be treated as verbatim copying in other respects. If the required texts for either cover are too voluminous to fit legibly, you should put the first ones listed (as many as fit reasonably) on the actual cover, and continue the rest onto adjacent pages. If you publish or distribute Opaque copies of the Document numbering more than 100, you must either include a machine-readable Transparent copy along with each Opaque copy, or state in or with each Opaque copy a computer -network location from which the general network -using public has access to download using public -standard network protocols a complete Transparent copy of the Document, free of added material. If you use the latter option, you must take reasonably prudent steps, when you begin distribution of Opaque copies in quantity, to ensure that this Transparent copy will remain thus accessible at the stated location until at least one year after the last time you distribute an Opaque copy (directly or through your agents or retailers) of that edition to the public. It is requested, but not required, that you contact the authors of the Document well before redistributing any large number of copies, to give them a chance to provide you with an updated version of the Document.

4. MODIFICATIONS

You may copy and distribute a Modified Version of the Document under the conditions of sections 2 and 3 above, provided that you release the Modified Version under precisely this License, with the Modified Version filling the role of the Do cument, thus licensing distribution and modification of the Modified Version to whoever possesses a copy of it. In addition, you must do these things in the Modified Version: A. Use in the Title Page (and on the covers, if any) a title distinct from that o f the Document, and from those of previous versions (which should, if there were any, be listed in the History section of the Document). You may use the same title as a previous version if the original publisher of that version gives permission. B. List on the Title Page, as authors, one or more persons or entities responsible for authorship of the modifications in the Modified Version, together with at least five of the principal authors of the Document (all of its principal authors, if it has fewer than f ive), unless they release you from this requirement. C. State on the Title page the name of the publisher of the Modified Version, as the publisher. D. Preserve all the copyright notices of the Document. E. Add an appropriate copyright notice for your modi fications adjacent to the other copyright notices. F. Include, immediately after the copyright notices, a license notice giving the public permission to use the Modified Version under the terms of this License, in the form shown in the Addendum below. G. Preserve in that license notice the full lists of Invariant Sections and required Cover Texts given in the Document's license notice. H. Include an unaltered copy of this License. I. Preserve the section Entitled "History", Preserve its Title, and add to it an item stating at least the title, year, new authors, and publisher of the Modified Version as given on the Title Page. If there is no section Entitled "History" in the Document, create one stating the title, year, authors, and publisher of the Document as given on its Title Page, then add an item describing the Modified Version as stated in the previous sentence. J. Preserve the network location, if any, given in the Document for public access to a Transparent copy of the Document, and likewise the netwo rk locations given in the Document for previous versions it was based on. These may be placed in the "History" section. You may omit a network location for a work that was published at least four years before the Document itself, or if the original publish er of the version it refers to gives permission. K. For any section Entitled "Acknowledgements" or "Dedications", Preserve the Title of the section, and preserve in the section all the substance and tone of each of the contributor acknowledgements and/or d edications given therein. L. Preserve all the Invariant Sections of the Document, unaltered in their text and in their titles. Section numbers or the equivalent are not considered part of the section titles. M. Delete any section Entitled "Endorsements". S uch a section may not be included in the Modified Version. N. Do not retitle any existing section to be Entitled "Endorsements" or to conflict in title with any Invariant Section. O. Preserve any Warranty Disclaimers. If the Modified Version includes new f ront-matter sections or appendices that qualify as Secondary Sections and contain no material copied from the Document, you may at your option designate some or all of these sections as invariant. To do this, add their titles to the list of Invariant Secti ons in the Modified Version's license notice. These titles must be distinct from any other section titles. You may add a section Entitled "Endorsements", provided it contains nothing but endorsements of your Modified Version by various parties --for example , statements of peer review or that the text has been approved by an organization as the authoritative definition of a standard. You may add a passage of up to five words as a Front -Cover Text, and a passage of up to 25 words as a Back -Cover Text, to the end of the list of Cover Texts in the Modified Version. Only one passage of Front-Cover Text and one of Back-Cover Text may be added by (or through arrangements made by) any one entity. If the Document already includes a cover text for the same cover, previously added by you or by arrangement made by the same entity you are acting on behalf of, you may not add another; but you may replace the old one, on explicit permission from the previous publisher that added the old one. The author(s) and publisher(s) of the Document do not by this License give permission to use their names for publicity for or to assert or imply endorsement of any Modified Version.

5. COMBINING DOCUMENTS

You may combine the Document with other documents released under this License, under the terms defined in section 4 above for modified versions, provided that you include in the combination all of the Invariant Sections of all of the original documents, unmodified, and list them all as Invariant Sections of your combined work in its lic ense notice, and that you preserve all their Warranty Disclaimers. The combined work need only contain one copy of this License, and multiple identical Invariant Sections may be replaced with a single copy. If there are multiple Invariant Sections with the same name but different contents, make the title of each such section unique by adding at the end of it, in parentheses, the name of the original author or publisher of that section if known, or else a unique number. Make the same adjustment to the sectio n titles in the list of Invariant Sections in the license notice of the combined work. In the combination, you must combine any sections Entitled "History" in the various original documents, forming one section Entitled "History"; likewise combine any sect ions Entitled "Acknowledgements", and any sections Entitled "Dedications". You must delete all sections Entitled "Endorsements".

6. COLLECTIONS OF DOCUMENTS

You may make a collection consisting of the Document and other documents released under this Lice nse, and replace the individual copies of this License in the various documents with a single copy that is included in the collection, provided that you follow the rules of this License for verbatim copying of each of the documents in all other respects. Y ou may extract a single document from such a collection, and distribute it individually under this License, provided you insert a copy of this License into the extracted document, and follow this License in all other respects regarding verbatim copying of that document.

7. AGGREGATION WITH INDEPENDENT WORKS

A compilation of the Document or its derivatives with other separate and independent documents or works, in or on a volume of a storage or distribution medium, is called an "aggregate" if the copyright resulting from the compilation is not used to limit the legal rights of the compilation's users beyond what the individual works permit. When the Document is included in an aggregate, this License does not apply to the other works in the aggregate which are not themselves derivative works of the Document. If the Cover Text requirement of section 3 is applicable to these copies of the Document, then if the Document is less than one half of the entire aggregate, the Document's Cover Texts may be placed on covers that bracket the Document within the aggregate, or the electronic equivalent of covers if the Document is in electronic form. Otherwise they must appear on printed covers that bracket the whole aggregate.

8. TRANSLATION

Translation is considered a k ind of modification, so you may distribute translations of the Document under the terms of section 4. Replacing Invariant Sections with translations requires special permission from their copyright holders, but you may include translations of some or all I nvariant Sections in addition to the original versions of these Invariant Sections. You may include a translation of this License, and all the license notices in the Document, and any Warranty Disclaimers, provided that you also include the original Englis h version of this License and the original versions of those notices and disclaimers. In case of a disagreement between the translation and the original version of this License or a notice or disclaimer, the original version will prevail. If a section in t he Document is Entitled "Acknowledgements", "Dedications", or "History", the requirement (section 4) to Preserve its Title (section 1) will typically require changing the actual title.

9. TERMINATION

You may not copy, modify, sublicense, or distribute th e Document except as expressly provided for under this License. Any other attempt to copy, modify, sublicense or distribute the Document is void, and will automatically terminate your rights under this License. However, parties who have received copies, or rights, from you under this License will not have their licenses terminated so long as such parties remain in full compliance.

10. FUTURE REVISIONS OF THIS LICENSE

The Free Software Foundation may publish new, revised versions of the GNU Free Documentat ion License from time to time. Such new versions will be similar in spirit to the present version, but may differ in detail to address new problems or concerns. See http://www.gnu.org/copyleft/. Each version of the License is given a distinguishing version number. If the Document specifies that a particular numbered version of this License "or any later version" applies to it, you have the option of following the terms and conditions either of that specified version or of any later version that has been pub lished (not as a draft) by the Free Software Foundation. If the Document does not specify a version number of this License, you may choose any version ever published (not as a draft) by the Free Software Foundation. ADDENDUM: How to use this License for yo ur documents To use this License in a document you have written, include a copy of the License in the document and put the following copyright and license notices just after the title page: Copyright (c) YEAR YOUR NAME. Permission is granted to copy, distr ibute and/or modify this document under the terms of the GNU Free Documentation License, Version 1.2 or any later version published by the Free Software Foundation; with no Invariant Sections, no Front -Cover Texts, and no Back-Cover Texts. A copy of the li cense is included in the section entitled "GNU Free Documentation License". If you have Invariant Sections, Front -Cover Texts and Back -Cover Texts, replace the "with...Texts." line with this: with the Invariant Sections being LIST THEIR TITLES, with the Fr ont-Cover Texts being LIST, and with the Back -Cover Texts being LIST. If you have Invariant Sections without Cover Texts, or some other combination of the three, merge those two alternatives to suit the situation. If your document contains nontrivial examp les of program code, we recommend releasing these examples in parallel under your choice of free software license, such as the GNU General Public License, to permit their use in free software.

FSC
www.fsc.org
MIX
Papier | Fördert
gute Waldnutzung
FSC® C083411

Kolibri 360 GmbH
Ferdinand-Jühlke-Straße 7
99095 Erfurt, Deutschland
produktsicherheit@kolibri360.de

Druck:
CPI Druckdienstleistungen GmbH
im Auftrag der
Kolibri 360 GmbH
Ferdinand-Jühlke-Str. 7
99095 Erfurt